Archives of Science

Challenges and Opportunities in the 21st Century

Edited by Polina E. Ilieva and Venkat Srinivasan

University of California
Center for Health Humanities
Department of Humanities and Social Sciences
UCSF (Box 0850)
490 Illinois Street, Floor 7
San Francisco, CA 94143-0850

www.UCHealthHumanitiesPress.com

This series is made possible by the generous support of the Dean of the School of Medicine at UCSF, the UCSF Library, and a Multicampus Research Program Grant from the University of California Office of the President. Grant ID MR-15-328363 and Grant ID M23PR5992.

Designed by Virtuoso Press
Cover illustration: Megha Ramachandra

ISBN (pbk): 979-8-9926888-3-2
https://doi.org/10.5072/FK29310W36

Printed in USA

Contents

Introduction

Polina E. Ilieva and Venkat Srinivasan

There is an increasing need to evaluate the preservation of, and access to, the process of science in the twenty-first century. Archives of contemporary science are challenged with long-term responsibilities of preserving scientific records produced since the post-World War II period. In addition, they also need to respond to the requirements for wider access, diversity in the composition of the archive, ethics of access, privacy and use, and a balance between care and efficiency in archival processing.

To respond to these challenges, the 4th Workshop on Scientific Archives was hosted by the UCSF Library Archives and Special Collections on June 5 and June 6, 2024, in San Francisco. The workshop's program was developed by members of the Committee on the Contemporary Archives of Science and Technology of the International Council on Archives Section on University and Research Institution Archives (ICA-SUV). This international gathering provided a unique platform for fostering collaboration, forging partnerships, and building interdisciplinary dialogue among archivists, researchers, museum professionals, data scientists, and other stakeholders from across the globe. By synergizing collective expertise and experiences, the aim was to chart a course toward a more robust, inclusive, and sustainable framework for scientific archiving. The 2024 workshop was the largest event since the inception of the series in 2016 (past workshops included the first in Heidelberg in 2016, the second in Washington D.C. in 2018, and the third that was held virtually from Hamburg in 2021), with more than 60 participants,18 paper presentations and numerous poster presentations spread over the two days. Attendees came from eight countries and ten states across the United States, with Dr. Catherine R. Lucey, Executive Vice Chancellor and Provost at the University of California, San Francisco (UCSF), providing the opening remarks.

The workshop (https://tiny.ucsf.edu/scientific) was distributed across six sessions that focused on open archives, collections as data, scientific museums and archives, ethics of preservation and access in medical archives, appraisal of scientific records, and silences and gaps in scientific archives Presentations emphasized the theme of open archives and open science, covering topics such as:

- Partnership between scientists and archivists in the appraisal of modern scientific records
- Barriers to creating open archives
- Silences and gaps in scientific archives
- Archives practices for increasing equity
- Scientific archives and digital humanities/digital health humanities
- AI and scientific archives
- Collections and archives as data
- Developing professional skills and knowledge base in scientific archives
- Collaborative science and Big Team science challenges for archiving
- Opportunities for engagement, teaching with, and exhibiting scientific collections

The workshop also included an "unconference"-styled participant-driven event at the end. Through collaboration, participants identified an agenda and discussed topics focused on prominent themes derived from input collected during the workshop. It successfully achieved its goal of sharing challenges, developing solutions, generating ideas, and building partnerships.

This volume includes twelve chapters, which draw from the authors' papers and presentations at the workshop.

*

There is increasing overlap of work done by institutions across the galleries-libraries-archives-museums (GLAM) sector. Archives often have a library, or are also associated with institutional libraries. Museums and collection centers often interface with their archival counterparts and may have a library open to the general public and professionals researcher. Software developers often find themselves having to consider modules on software that will adapt a 'museum' cataloguing system for an exhibition; or an 'archive' cataloguing system for a collection center. Libraries often house rare material without necessarily having a separate archive or special collections. And of course, every institution aims to have an exhibition space that is open to the public. The first two chapters in this volume discuss the relation between museums and archives from distinctly different perspectives

In their chapter titled, "Scientific Archives and Museum Collections: An Integrated Vision," Patrícia Costa, Milena Carvalho and Susana Martins reflect on the collection of scientific instruments at the Museum of the Instituto Superior de Engenharia do Porto, Polytechnic of Porto, Portugal. They illustrate how these historical instruments act as "infocommunicational" objects that help document the evolution of industrial education in Porto. They investigate the integration of scientific archives and museums, including how digital tools have improved access to these museum objects.

Maya Naunton and Kiana Clark discuss a very different kind of scientific museum and archival objects in their chapters on "Collaboration Between Archivists and Scientists at the American Museum of Natural History" and "Restrictions and Data in the Vertebrate Paleontology Archive." The organic growth at the 130-year-old Vertebrate Paleontology Department at the American Museum of Natural History (AMNH) has posed both a challenge and an opportunity, and they illustrate work on three different axes at the intersection of their archives and collections database. Close collaboration between scientific staff and the archivists has led to better organization and descriptions, and also ensured scientists understand the role of archives. They discuss plans to develop connections between the specimen records in their collections database and archival material. And they illustrate the balance needed between open access to the archives, and placing necessary restrictions for personal data privacy as well as protecting the location of the field sites for fossils.

Nowhere is ethics of preservation and access as critical, perhaps, as in medical archives. The volume includes two chapters on this theme. In their chapter, "Beyond the Walls and Changing the Space of Archive Through the Lens of The Adler Museum of Medicine in South Africa," Samuel Umoh Uwem (University of Hradec Kralove, Czech Republic) and Adetola Elizabeth Umoh (University of Autonomous Barcelona, Spain) examine the digitization practices at the archives at the Adler Museum. They review the choices to "include or exclude medical archives, and how they incorporate these insights into their digital archives' design, functionality, and affordance." The paper draws on archival sources, interviews and secondary reference material to reflect on both the methods employed, and on the concerns around a "coming digital dark age," when archives are lost due to rapid digitization without sufficient regard to preservation.

Michaela Clark (Heinrich Heine University in Düsseldorf and University of Cape Town Pathology Learning Centre) gives an overview of how over 6000 clinical settler-colonial "orphaned" and decontextualized photographs preserved at a medical school were restored with archival and research value in the University of Cape Town in South Africa. In their chapter, "Curating the Clinical: Visual and Material Methods for a Photographic Collection at Cape Town's Medical School," they discuss how the photographs, which were produced between 1920 and 1967 and depict the bodies of local patients, had no records about their origin and brought up numerous ethical concerns including queries about patient consent. The photographs were recontextualized as both visual and material records and subject to provenance research. Clark discusses the process of this mode of curation that enabled the 'photo-objects' to serve as a site for writing histories of "visual and material medical culture in postcolonial contexts."

Understanding and displaying provenance is one of the fundamental principles of archival processing, and leads directly to the archivist's appraisal of scientific records, which is the theme for four chapters in this volume. Like in other themes, these chapters offer a distinct variety in appraisal methodologies and experiments at various institutions, from the challenges of appraising digital data to reviewing appraisal theory as a mode of evaluation, and from appraising records related to decades of robotics research to the modern use of artificial intelligence across disciplines.

Tommy Keswick and Mariella Soprano discuss their innovative experiment at the Caltech Library in their chapter, "AI-Enhanced Appraisal and Processing of a Hybrid Collection." With the increasing presence of both physical and digital records in vast quantities, archives find it difficult to appraise across media in an efficient way. For their case study, they discuss the processing at Caltech Archives of the papers of Robert H. Grubbs, a renowned chemist. They show how they use various digital tools across the workflow, including modern text recognition and machine learning algorithms. Through their novel experiments with this collection, they illustrate some of the first attempts at an AI-enhanced workflow toward various aspects of the process, including eliminating duplication of archival data across physical and digital media, enhancing descriptions, and arrangement of both digital records and photographed documents in an "intellectually unified finding aid."

A similar appraisal challenge in response to modern needs is discussed in the chapter, "Reappraising "Appraising the Records of Modern Science and Technology" Helping Archivists Document 21st Century Science" by Bethany G. Anderson, Polina E. Ilieva, and Jordon Steele from the United States. They revisit the 1985 guidelines for appraising science and technology records published by Joan K. Haas, Helen Willa Samuels, and Barbara Trippel Simmons. Their chapter discusses the critical place of the guide in shaping decades of archival work across institutions and its holistic approach to appraisal theory including functional analysis and documentation strategy. But science itself has changed considerably since 1985, with the sheer volume of digital data, as well as a robust open size movement. This brings into focus a need to review the guide and to also conceptual appraisal questions relating to the impact of computers and the Internet, how the site of "doing" science has evolved over decades, the place of diversity, equity, inclusion and accessibility in science, and the very nature of a science archive.

Fernanda Ribeiro and Armando Malheiro da Silva from the University of Porto, Portugal, offer a different structure to appraisal theory in itself in their chapter, "The Appraisal of Scientific Documentation/Information as a Methodological Operation: Application Criteria and Parameters." In their chapter, they discuss how evaluation in information services is done through evaluation of services, of information retrieval and of appraisal of information flow. Their work positions appraisal as a mode of evaluation, and as a methodological operation applicable to information in "any production use content." They argue that appraisal, therefore, does not "have an end in itself, disconnected from a broader method that associates the comprehensive or scientific aspect with the applicational or technical one." Their chapter discusses the appraisal model developed at the University of Porto, which, they write, brings in a life cycle approach to information and can be applied to any type of information within the framework of information sciences.

The work of Julia Corrin and Kathleen Donahoe from the Carnegie

Mellon University (United States), interestingly, weaves together questions of both appraisal and of working with subject experts in making these decisions, echoing similar archival questions raised in the chapter by Maya Naunton and Kiana Clark from the American Museum of Natural History, United States. In their chapter, "The Robotics Project: Collecting Complex Science at Carnegie Mellon," Corrin and Donahoe discuss the collaboration between their University Libraries and the School of Computer Science, drawing on technical expertise to develop best practices for archiving robotic materials, especially given their varied form and content. The Carnegie Mellon University Archive created a new Digital Robotics Archive that serves as the digital home for some of the appraised material. The university's long tradition in artificial intelligence (AI) from work on the first AI program in the 1950s and the Field Robotics Center in the 1980s, to continued work in commercialization of robotics in the 1990s, all point to a growing need to integrate and rethink archival practices in this discipline.

Perhaps as fundamental a challenge in archives today is the simple question: whose archives? Issues relating to the composition of the archive and access to aspects of its collections are sprinkled through all the chapters. They come into sharp focus in three chapters that look at silences and gaps in scientific archives.

Deepika S and Anjali Ramachandran from the Archives, National Centre for Biological Sciences, India (Archives at NCBS), discuss the struggle and opportunity in bringing in histories of marginalized groups in their chapter, "Building an Inclusive Science Archive in India." Their work reflects on a rethink in their archive's collection mandate, by also examining the limitations of certain aspects of archival theory in highlighting the presence of under-represented histories. A science archive like theirs, they argue, has to contend with doing a lot at the same time: broadly preserve scientific heritage including those that come from privilege but don't have ready archives to send their material, identify and preserve archival material from under-represented communities, and do all of that with limited resources. In their chapter, they discuss their work on using feminist standpoint appraisal in their approach to archiving, "to include the people and communities that are archived and involved in the process of archiving."

Traditional archival finding aids often hide histories of marginalized communities, and this is echoed in the work of David Ragnar Nelson (American Philosophical Society) and Serenity Sutherland (SUNY Oswego). In their chapter,

"Connections, Collaborations, Collections: Using Data Visualization to Rethink Access to Scientific Archives," they offer a novel approach to discovery of these histories through data visualization tools. With the help of an interactive network visualization project titled "Visualizing Women in Science," users can find information about women scientists in the archives in ways that would have otherwise stayed hidden in the traditional finding aid. They also discuss the development of a prototype subject guide for all the collections at the American Philosophical Society, and effectively offer new ways to find material. This reframing, they argue, helps in "confronting and remediating archival silences."

Finally, Dan Kabella, Kelly Ray Knight, and Dorie Apollonio from the University of California, San Francisco bring in a distinct perspective on archival silences through their work on the opioid industry. As outlined in their chapter title, "Extending the Commercial Determinants of Health: Lessons from the Opioid Industry Archive," they extend an existing framework on the influence of corporate practices in health outcomes by analyzing the Opioid Industry Documents Archive (OIDA), a public repository of internal industry documents. This includes documents on how industry hid the harmful effects of opioids and framed pain management and addiction medication in the United States. Kabella, Knight and Apollonio's work specifically sees the OIDA as much more than a repository and as a "generative infrastructure" that can help illustrate how "knowledge about harm is produced, legitimated, and contested across institutional, community, and epistemic boundaries." Their work moves away from seeing corporate influence in medicine as being disruptive and specifically argues that this corporate power is both embedded and builds the infrastructure of clinical practice and public health governance and discourse.

The chapters in this volume touch upon such pressing challenges at science archives across the world today: linkages between museums, collections and archives, issues around ethics of preservation and access, fundamental challenges in appraisal, and archival silences engendered by traditional discovery tools. This volume offers perspectives from various corners of the world, all dealing with similar issues, and with various experiments to address the challenges of the 21st century science archive. We hope the reader finds much for reflection through this volume, and that it sets the stage for further discussion on the shape of a science archive.

Acknowledgments

The success of the 4th Workshop on Scientific Archives would not have been possible without the tireless efforts and dedication of numerous individuals. We would like to extend our heartfelt thanks to UCSF Archives and Special Collections team members for their invaluable contributions: Peggy Tran-Le, Lupe Samano, Edith Escobedo, Lisa Nguyen, Emma James whose dedication made this event a resounding success.

Special thanks are extended to the UCSF University Librarian, Chris Shaffer, for his support.

Additionally, we would like to recognize the entire organizing committee for their collaborative spirit. Your collective expertise, hard work, and enthusiasm have been truly inspiring and instrumental in fostering an engaging and productive event:

- Bethany G. Anderson, University of Illinois Urbana-Champaign, United States
- Jean Marie Deken, SLAC Archives, History & Records Office (AHRO) and Research Library, Stanford, United States
- Polina E. Ilieva, associate university librarian for collections and UCSF archivist, United States
- Anne-Flore Laloë, European Organization for the Exploitation of Meteorological Satellites (EUMETSAT), Germany
- Rebekah Kim, California Academy of Sciences Library and Archive, United States
- Ana Margarida Dias da Silva, Archives of the University of Coimbra's Department of Life Sciences, Center for the History of Society and Culture, Portugal
- Melanie Mueller, American Institute of Physics, United States
- Laura Outterside, records manager, European X-Ray Free-Electron Laser Facility, Germany

- Patrick Shea, chief curator of archives and manuscripts for the Othmer Library of Chemical History at the Science History Institute, United States
- Venkat Srinivasan, head, Archives at the National Centre for Biological Sciences (Archives at NCBS), Bengaluru, India

The publication of this book was made possible through generous support by the UC Health Humanities Press and editor-in-chief, Brian Dolan, PhD.

1

Scientific Archives and Museum Collections: An Integrated Vision

Patrícia Costa [1]; Milena Carvalho [2]; Susana Martins [3]

Abstract

Instituto Politécnico do Porto (IPP), a prominent public higher education institution, has established a strategic plan encompassing various fields including arts, health, and engineering. This paper examines the historical museum collections associated with IPP's engineering and accounting institutes, showcasing their significance in documenting the evolution of industrial education in Porto. Through an exploratory study, it investigates the integration of scientific archives and museums, focusing on scientific instruments as infocommunicational objects. By employing methodologies from literature review and interpretive research, the analysis underscores the importance of these artifacts in fostering interdisciplinary education and cultural memory. Furthermore, the study highlights the role of digital innovations in enhancing access and engagement with museum collections. It ultimately demonstrates how museum objects serve not only as historical artifacts but also as educational tools, bridging past practices with contemporary teaching methodologies.

[1] ISEP / P.PORTO, Rua Dr. António Bernardino de Almeida, 431. 4249-015 Porto, Portugal

[2] CITCEM, FLUP, U. Porto; CEOS.PP / ISCAP / P.PORTO, Rua Jaime Lopes Amorim s/n, 4465-004, Matosinhos, Portugal. milenacarvalho@iscap.ipp.pt

[3] CEOS.PP / ISCAP / P.PORTO, Rua Jaime Lopes Amorim s/n, 4465-004, Matosinhos, Portugal; CITCEM, FLUP, U. Porto

Introduction

P. Porto is a referential public higher education institution. Beginning its activity in 1985, it´s today the fifth largest institution among universities. It also ranks fourth among first-choice higher education institutions.

P. Porto Strategic Plan has defined the following strategic action areas:

- Arts and Humanities;
- Education;
- Engineering, Manufacturing and Construction Industries;
- Health and Social Protection;
- Science, Mathematics and Computer Science;
- Services;
- Social Sciences, Business and Law.

Both the higher engineering institute of Porto, founded in 1852 as the Industrial School of Porto, and the Higher Institute of Accounting for Administration of Porto, founded in 1886 as the Commercial Institute of Porto, both schools of the Polytechnic of Porto, have a museum collection associated with the teaching activity.

The Museum of the Higher Engineering Institute of Porto (ISEP), Polytechnic of Porto, houses an important collection of scientific instruments of great scientific and pedagogical quality that clearly reflects a great deal of scientific knowledge about the circulation of new ideas and techniques, which allowed a country like Portugal to develop industrial education and experiential training, with reference to the most technically developed countries, as France, England and Germany.

Due to its characteristics and underlying history, this museum collection is of considerable historical and scientific interest for the study of the different teaching methods practiced in industrial education in the city of Porto, even when compared to today's standards. In this way, we consider the museological object to be a document, since it is part of a chain of contexts, links and connections, and is an informational envelope.

The work presented here is an explanatory study as it aims to provide in-depth knowledge of the subject in question, based on a literature review and documentary research.

Methodology

The study presented is based on an exploratory study, seeking to investigate a problem or topic that has not yet been fully characterized. Exploratory study is not intended to produce definite evidence, but it helps understanding the topic. Exploratory studies often use interpretive research methodologies, but they demand a confirmatory literature review (Dash, 2019).

Further expanding the discussion, Green et al. (2022) address the complexities inherent in case study research and its implications for causal inference. They reference Gerring's definition of case study research as an intensive examination of a single unit to gain insights into broader classes of similar units. This perspective is particularly valuable in elucidating deterministic causes and understanding the mechanisms that underpin causal relationships. By categorizing types of claims related to causality, the authors provide a structured approach to evaluating evidence and the conditions that contribute to complex causal relationships. This work highlights the epistemological challenges, and the diverse methodologies employed in explanatory studies, reinforcing the importance of clarity in research design and the articulation of causal claims.

Shmueli (2011) argues that explanatory frameworks is essential for advancing scientific understanding.

T. Mueller et al. (2019) highlights the growing importance of distinct explanatory frameworks. Their meta-review integrates essential concepts from multiple fields, including philosophy and psychology, to examine ideas like causal reasoning and hypothesis development. The writers emphasize the importance of explanation in scientific reasoning, arguing that comprehending the mechanisms underlying phenomena is essential for empirical research. This analysis shows that explanatory studies focus not only on results but also on the processes that lead to those results, thereby enhancing the conversation about explanation in modern research.

The literature review serves as a crucial foundation for understanding the characteristics of explanatory studies and their significance in advancing knowledge within a specific domain. In this context, Amobonye et al. (2024)

articulate that the primary objective of literature reviews is to furnish direction and information conducive to the enhancement of scientific knowledge. They emphasize that conducting a thorough literature review not only demonstrates the author's familiarity with the field but also encompasses a comprehensive exploration of its historical context, theoretical frameworks, key variables, terminologies, phenomena, and research methodologies.

Furthermore, the authors underscore the pivotal role of review articles in the scientific community, highlighting their capacity to convey insights while simultaneously identifying existing knowledge gaps that warrant further exploration in future research endeavors. This identification of gaps is particularly significant in explanatory studies, where understanding the underlying causes and relationships between variables is essential. The authors note that review articles often garner more citations compared to original research papers, indicating their influential nature and the broader reach they provide for researchers. This trend suggests that literature reviews are not merely supplementary; rather, they are integral to the scholarly discourse and can greatly enhance a researcher's visibility and impact within their field.

In addition, the references included in a literature review serve as a valuable resource for readers, allowing them to delve deeper into the subject matter and broaden their understanding. This interconnectedness of knowledge is vital for explanatory studies, which rely on a robust framework of existing research to elucidate complex phenomena. Through the lens of the insights provided by Amobonye et al. (2024), it becomes evident that literature reviews are instrumental in shaping the landscape of scientific inquiry, particularly in the context of explanatory studies, where clarity and depth of understanding are paramount.

1. Exploring the integration of scientific archives and scientific museums: viewing scientific objects as documents

This works is based on the premise that scientific archives and museums within higher education institutions hold significant cultural, pedagogical, and epistemological value. The framework presented explores the intersection of museology and information science, the role of scientific instruments as both educational tools and museum artifacts, and the conceptualization of museum objects as infocommunicational objects.

The museum collection cannot be dissociated from an important set of documents that proves all the school's administrative activity, as well as the record of scientific activity carried out in the different subjects at the school between 1852 and 1974 - the Historical Archive.

All these elements should be approached holistically, thus forming and consolidating the basis of the information system (IS) with a focus on information potential, contributing to the efficient management of all the heritage information produced in a museum context.

The holistic approach is primarily since the process of gathering, storing, processing and providing access to infocommunicational objects (data recipients) relevant to the mission of the institution and to its' target audience are present in both museums and archives.

It must also include the availability of information associated with the historical archive and documentation center and other data from the institution that is considered relevant, allowing for perfect and transparent articulation between all aspects.

Museology and Information Science have an important connection as they both focus on knowledge preservation, documentation, and dissemination. Museology benefits from information science because it provides methods for organizing, storing, and retrieving data on things, whereas information science benefits from museology's emphasis on the contextual and narrative importance of artifacts.

We can create the biography of objects through a variety of sources and approaches.

The biographical study of objects, developed by Karin Dannehl, combines two concepts:

1. The life cycle
2. The study of life cycle assessment

The integration of scientific archives and museums is a dynamic area of study that highlights the intersection of data management, preservation needs, and heritage management. The literature in this field reveals significant insights into how viewing scientific museum objects as documents can enhance information systems and contribute to the effective stewardship of scientific data.

We can easily identify areas in which this intersection occurs: Knowledge Organization: In museums within higher education institutions, objects are not just displayed; they are cataloged with metadata that includes provenance, use, and historical context. This organization enables effective information retrieval and supports research and education (Buckland, 2017); Digital Museology: Advances in information science, such as database systems and digital curation, have transformed how museums document and present their collections. Digitized archives provide global accessibility and allow researchers to engage with museum objects remotely (Parry et al., 2020); Epistemological Synergy: Both fields contribute to the creation of what Floridi (2011) refers to as the "infosphere," where physical and digital knowledge artifacts coexist and are interlinked. **A clear real example is that in ISEP**, historical scientific instruments housed in ISEPs´ museum are documented not only as artifacts but also as carriers of pedagogical traditions and scientific breakthroughs.

The transition of scientific instruments from functional tools to museum objects exemplifies the interplay between the history of science, pedagogy, and museology. These instruments serve as tangible records of educational practices and the evolution of scientific inquiry. We can regard to these o**bjects as Documents:** Scientific instruments in higher education museums embody documentation of teaching methods, technological advancements, and the operational history of institutions. They are "documentary objects" (Briet, 1951) that carry information about their context of use; **Educational Legacy:** Instruments like early microscopes or physics apparatuses are tied to the curriculum and teaching methodologies of their time. Accompanying manuals, notebooks, and other pedagogical materials enhance their interpretive value (Forgan, 2005); **Preservation of Functionality:** Many university museums prioritize preserving the operability of these instruments, enabling demonstrations that bridge historical and contemporary scientific methods. In some institutions like Harvard or the University of Oxford, collections of 18th and 19th-century apparatuses are used to illustrate historical teaching practices, connecting past and present educational paradigms.

Akmon et al. (2011) discuss the necessity for archivists to adapt their understanding of data management within the context of data-intensive scientific environments. They argue that traditional assumptions about archival practices, which may have been effective in more bureaucratic settings, often fail to hold true in the rapidly evolving landscape of digital data. This observation underscores the importance of updating archival practices to remain relevant and effective in supporting scientific data curation. The authors emphasize that by aligning their skills with the unique needs of scientists, archivists can play a crucial role in facilitating better data management and preservation, thereby enhancing the overall integration of scientific archives and museums.

Downs & Chen (2012) explore the challenges posed by the digital revolution in their examination of sustainable stewardship of digital collections. They highlight the exponential growth in the volume and variety of scientific data, which has outpaced the capacity of traditional institutions such as libraries and museums to manage and preserve these resources effectively. The authors advocate for the establishment of collaborative partnerships across different types of organizations to address the pressing need for long-term data stewardship. They point out that developing viable economic models and robust spatial data infrastructures is essential for ensuring the sustainability of these initiatives. Their work illustrates the critical role that scientific data stewardship plays not only in preserving historical data but also in enabling ongoing scientific inquiry and progress.

Together, these articles provide a comprehensive overview of the evolving landscape of scientific data management and highlight the imperative for archivists and institutions to adapt their practices in response to the challenges posed by digital data. The integration of scientific archives and museums, viewed through the lens of these scholarly contributions, reveals a promising pathway for enhancing information systems and heritage management in an increasingly data-driven world and look at the infocommunicational object / artifact as a version of the human thinking.

According to Choudry & Hanisch (2009) the importance of preservation is a concept very present as the Data Conservancy (DC) was funded by the US National Science Foundation's Data Net program, aiming to create a sustainable system for the curation and preservation of interdisciplinary scientific data. Led by the Sheridan Libraries at Johns Hopkins University, the initiative focuses on developing a comprehensive data curation strategy that integrates infrastructure research, computer science, and information science. By engaging with various scientific domains, including astronomy, life sciences, and social sciences, DC seeks to improve data accessibility and interoperability across disciplines. Key components include establishing a common data framework, addressing sustainability concerns, and enhancing the role of libraries in digital data stewardship. The project emphasizes collaboration among researchers, libraries, and educational institutions to foster a new generation of data scientists.

There seems to be accordance regarding the emphasizes that data curation is essential for addressing grand research challenges, rather than an end. A key focus is on creating a cross-disciplinary data framework that allows for effective data discovery and interoperability across diverse scientific fields and sustainable data curation requires long-term investment and an adaptable approach that considers the unique practices of various scientific disciplines.

Chourdy & Hanisch (2009) state that this initiative highlights the importance of training and education in developing a workforce skilled in data preservation and curation and that the collaborative model fosters interdisciplinary partnerships, which are crucial for tackling complex scientific questions that span traditional boundaries and essential to develop a sustainable system for the curation and preservation of interdisciplinary scientific data to address significant research challenges. The authors assert that sustainability is crucial because it ensures that data preservation efforts can continue over the long term, requiring both upfront and ongoing investments to maintain operational capabilities. Information services such as libraries and archives are positioned as leaders in data curation efforts, leveraging their expertise in object-information stewardship to enhance data management and accessibility in the digital age.

2. Infocommunicational objects and the mindfact / artifact approach

The exploration of infocommunicational objects through the lens of the mindfact-artifact approach reveals a complex interplay between mental representations and material manifestations. Authors like Kallinikos (2003), state that the distinction between context-free and context-embedded knowledge is most noticeable in bureaucratic environments, when the materiality of digital technologies is subordinated to symbolic organizing. The author underlines that the core of representation is inextricably linked to its physicality, implying that digital artifacts constitute a representational worldview that questions traditional conceptions of worldly reference. We add that this vision encompasses any infocommunicational object, even non-digital ones.

Greenhill and Fletcher (2009) argue that the meanings of artifacts are not fixed but are continuously negotiated and redefined through social interactions and temporal contexts.

Furthering this discourse, O'Raghallaigh, McCarthy, and Adam (2018) introduce the concept of sociomateriality, advocating for an object-inspired approach that emphasizes the experiential and imaginative engagement with objects, particularly technological ones. Their methodology encourages a reconnection of objects with their real-world contexts, fostering a deeper understanding of the production, usage, and meanings of these artifacts. This perspective not only enriches the theoretical foundations of information systems (IS) research but also highlights the intricate relationships between social and material dimensions in the study of infocommunicational objects.

There is a clear dynamic interplay between the mental constructions of infocommunicational items and their tangible counterparts, which sheds light on how these linkages impact user interactions and social experiences.

The "mind-fact"/artifact approach can be used to analyze informational items. The "mindfact" dimension refers to the mental image and/or emotional representation of information that is ready for materialization, as the artifact is the material and functional result that was idealized and produced from a mental image (Silva, 2013). The emphasis on information transmitted results in a more holistic approach to infocommunicational objects, taking into account context, relations, organization, and replication.

It´s assumed a vision based on the post-custodial, informational and scientific paradigm, and a transdisciplinary approach to Information Science. Information is a study object that can be defined as an organized set of codified representations that are directly related to social activities and can be recorded in any material medium (Silva & Ribeiro, 2002). This relates directly to Buckland's (1991) triple structure of the concept of information, which presents it in a cyclical and dynamic way:

- Information-as-Process: the act of informing;
- Information-as-Knowledge: which is perceived from the process, intangible, personal, subjective, conceptual;
- Information-as-Thing: referring to objects that can be informative, tangible; something expressed, described or represented in some physical way.

Museum objects in scientific archives are increasingly recognized as infocommunicational objects, e.g., entities that convey information and facilitate communication across time and disciplines. These objects have Infocommunicational Significance: according to Greenhill (1992), museum objects are repositories of encoded knowledge, functioning as communicative artifacts. In higher education museums, these objects stimulate inquiry and critical thinking among students and researchers; embody the Mindfact-Artifact Continuum: the concept of "mindfact" refers to the intellectual constructs derived from human interaction with objects (Latour, 1999). Museum artifacts are artifacts saturated with mindfacts, bridging material culture with cognitive processes. According to Carvalho, Martins & Pinto (2023), the mindfact represents the mental and emotional constructs encoded in objects, while the artifact is its physical manifestation. This binomial is crucial in interpreting museum objects as cognitive and communicative entities; Interdisciplinary Pedagogy: Viewing museum objects through this lens fosters interdisciplinary approaches, integrating history, science, and information studies; Informational Heritage: Carvalho et al. (2023) argue that the documentation of objects in museum settings bridges information science and museology by materializing abstract ideas into accessible educational and research tools. For example, scientific objects serve as case studies for discussions on epistemology, design thinking, and data visualization. Another example we can present is that a 19th-century astrolabe in a university museum not only represents historical astronomy but also serves as a medium to explore the mathematical and cultural context of its use; another aspect is the **Preservation and Functionality:** Research by Martins, Carvalho & Pinto (2023) emphasizes maintaining the functionality of historical scientific instruments in museums, arguing this approach enriches educational narratives and supports interdisciplinary learning; and **Sustainability of Cultural Memory:** Ribeiro & Silva (2023) and Martins, Carvalho & Pinto (2023) address the importance of sustainable management practices in museums to preserve cultural and scientific heritage for future generations.

In examining the evolving landscape of Information Science, it becomes apparent that the discipline grapples with the multifaceted nature of information itself, which is inherently complex and resistant to singular definitions. Carlin (2009) highlights this challenge, noting that the ambiguity surrounding the term "information" contributes to a division of labor across various library and information related fields. Carlin (2009) argues for a shift in focus from merely defining information to developing concepts that better capture its essence. This conceptual foundation is crucial for advancing theoretical work within the discipline, particularly as it relates to understanding the social implications of information and its cognitive dimensions.

Building upon this notion, Hemmer et al. (2014) further explore the intricacies of human interaction with information in computer-mediated contexts. Their review emphasizes the need for a cohesive framework that synthesizes existing research on human information behavior, using the term "Human Computer-based Information Behavior" (HCIB). This perspective broadens the understanding of information interactions, acknowledging the role of technology while also recognizing scenarios where such technology may be absent. Their work illustrates the importance of a pluralistic approach to information, which resonates with Carlin (2009) for conceptual clarity and coherence within the field.

In a more critical posture, McBride (2018) challenges the classification of Information Systems as a science, positing instead that it aligns more closely with the humanities. He argues that the study of information systems encapsulates the complex interplay of human creativity, relationships, and societal dynamics. McBride (2018) advises against taking an overly quantitative approach that ignores the complex, emergent phenomena inherent in human interactions with information systems. He has reservations about the discipline's fragmentation, which risks limiting it to simplistic, formulaic research that lack the creativity and diversity fundamental for comprehending complicated human processes and interactions.

Regarding the nature of information, the significance of human behavior in information interactions, including those with infocommunicational objects, and the broader implications of defining the discipline itself, a transdisciplinary approach is required to embrace the complexities of information as both a process and a product of human activity, reinforcing the need for a more nuanced understanding of the field's foundational concepts.

More recent approaches

Museology and information science continue to evolve with innovations in digital curation, metadata standards, and interactive technologies. Concerning Digital Innovations in Museum Practices: Marques and Sousa (2024) emphasize that digital systems and interactive tools have transformed the traditional museological role by improving public engagement and enabling multi-disciplinary data integration in university museums. These systems facilitate dynamic interpretation and access to collections.

Holistic approach is key to create informational and documentation systems. Chaleplioglou (2024) explores how information science methodologies are applied to museological objects to create integrated documentation systems that unify library, archive, and museum data. This convergence supports cross-disciplinary research in academic contexts. Examples include digitized catalogs that connect museum artifacts to historical academic records and virtual platforms enabling researchers to analyze interconnected datasets from diverse sources.

Recent studies reinforce the dual role of scientific instruments in education and as museum artifacts, emphasizing their documentary and interpretative potential. Montenegro et al. (2024) highlight that objects such as telescopes and laboratory tools are "infocommunicational artifacts," storing layers of historical and educational significance. These instruments enable a tangible connection to past teaching practices and Forgan (2023) call on the Documentation and Functionality, more specifically by preserving the functional aspects of scientific artifacts remains a priority. This approach is critical for demonstrating historical methodologies in modern educational settings. Forgan states as examples the restoration projects for historic apparatuses in university museums and the use of artifacts to re-enact historical experiments for pedagogy.

Regarding Infocommunicational Objects and the Mindfact-Artifact Connection, museum objects are infocommunicational entities, vital for bridging physical artifacts with digital and intellectual domains. Indeed, they embody a Cognitive and Narrative Integration: Greenhill's theory has been expanded by studies exploring how museum artifacts serve as mediators of complex narratives, combining physical presence with digital metadata (Shelegina, 2024) and Enhance Accessibility and Communication: Digital tools and AI technologies improve the accessibility of museum collections, transforming them into active participants in global scholarly discourse. Examples are AI-enhanced museum catalogs enabling multilingual and multidisciplinary access and metadata-rich object profiles that contextualize artifacts within broader academic narratives.

Conclusion

Museums and scientific archives in higher education institutions exemplify the synergy between museology and information science. Scientific instruments, as both historical artifacts and documentary objects, illuminate the pedagogical and operational history of institutions. Furthermore, the treatment of museum objects as infocommunicational entities emphasizes their role in fostering interdisciplinary education and research. As for practical applications we highlight the creation of Interactive Documentation Systems. Carvalho, Martins, Castro, & Santos (2023) propose digital platforms that integrate metadata about scientific and cultural objects, enhancing accessibility for researchers and educators and Global Engagement strategy that encompasses the digitization of collections, facilitate broader engagement for informational institutions, enabling the connecting of local cultural heritage to global society.

Recent research provides a robust theoretical grounding for the integrated vision of museology and information science as they emphasize concepts such as the mindfact-artifact relationship and the preservation of cultural and scientific heritage and highlight innovations and methodologies that connect historical artifacts, digital practices, and interdisciplinary knowledge in higher education museums.

This work highlights the evolving role of information science in enhancing museological practices and preserving the intellectual and cultural significance of scientific artifacts.

Acknowledgments

This texte is financed by Portuguese national funds through FCT – Fundação para a Ciência e Tecnologia, under the project UIDB/05422/2020;

This text is financed by National Funds through FCT – Foundation for Science and Technology, within the scope of the project UIDB/04059/2020.

References

Akmon, D., Zimmerman, A. S., Daniels, M., & Hedstrom, M. (2011). *The Application of Archival Concepts to a Data-Intensive Environment: Working with Scientists to Understand Data Management and Preservation Needs*. https://doi.org/10.1007/s10502-011-9151-4

Amobonye, A., Lalung, J., Mheta, G., & Pillai, S. (2024). Writing a Scientific Review Article: Comprehensive Insights for Beginners. *The Scientific World Journal, 2024*(1), 7822269. https://doi.org/10.1155/2024/7822269

Bennett, J. (2005). Museums and the History of Science: Practitioner's Postscript. *Isis, 96*(4), 602–608. https://doi.org/10.1086/498596

Buckland, M. (2017). *Information and Society*. MIT Press. https://mitpress.mit.edu/9780262533386/information-and-society/

Carlin, A. P. (2009). Segmented Professions: Further Considerations of Theory and Practice in LIS and Librarianship. *Library Philosophy and Practice*. https://www.semanticscholar.org/paper/Segmented-Professions%3A-Further-Considerations-of-in-Carlin/7c1dfe40bb086feb34d46bea9f07034d878a7d43

Carvalho, M., Martins, S., Castro, M. J., Santos, E., & de Carvalho, A. B. S. (2023). Informational Heritage and the Relation Between Information Science Museology: Information Professional Performance Projects. Em J. V. Carvalho, A. Abreu, P. Liberato, & A. Peña (Eds.), *Advances in Tourism, Technology and Systems: Selected Papers from ICOTTS 2022, Volume 1* (pp. 533–540). Springer Nature. https://doi.org/10.1007/978-981-99-0337-5_44

Carvalho, M., Martins, S., & Pinto, C. (2023). Landscape and cultural heritage: Object and information. *Heliyon*, *9*(10), e20395. https://doi.org/10.1016/j.heliyon.2023.e20395

Chaleplioglou, A. (2024). JIIM Bibliographic Guide 2018-2023. *Journal of Integrated Information Management*, *9*(1 Suppl.), Artigo 1 Suppl.

Dash, P. (2019). *Analysis of Literature Review in Cases of Exploratory Research* (SSRN Scholarly Paper 3555628). Social Science Research Network. https://doi.org/10.2139/ssrn.3555628

Downs, R. (2012). *Towards Sustainable Stewardship of Digital Collections of Scientific Data.* https://www.academia.edu/70051928/Towards_Sustainable_Stewardship_of_Digital_Collections_of_Scientific_Data

Floridi, L. (2013). *The Philosophy of Information* (Reprint edition). Oxford University Press.

Green, J., Hanckel, B., Petticrew, M., Paparini, S., & Shaw, S. (2022). Case study research and causal inference. *BMC Medical Research Methodology*, *22*(1), 307. https://doi.org/10.1186/s12874-022-01790-8

Greenhill, A., & Fletcher, G. (2009). The Polysemy of Human–Computer Interaction. Em H. Isomäki & P. Saariluoma (Eds.), *Future Interaction Design II* (pp. 175–190). Springer. https://doi.org/10.1007/978-1-84800-385-9_9

Greenhill, E. H. (1992). *Museums and the Shaping of Knowledge.* Routledge & CRC Press. https://www.routledge.com/Museums-and-the-Shaping-of-Knowledge/HooperGreenhill/p/book/9780415070317

Hanisch, R., & Choudhury, S. (2009). *The Data Conservancy: Building a Sustainable System for Interdisciplinary Scientific Data Curation and Preservation.* http://jhir.library.jhu.edu/handle/1774.2/34018

Hemmer, E., Heinzl, A., & Leidner, D. (2014). Foregrounding the "I" in IS Research: A Plea for Research on Computer-mediated Human Information Behaviour. Em *Working Paper Series in Information Systems* (Vols. 14–01) [Working paper]. https://ub-madoc.bib.uni-mannheim.de/35892

Kallinikos, J. (2003). Mediated Action and Representation on the Vicissitudes of Human Signification. *Homo Oeconomicus*, *19*, 607–622.

Martins, S., Carvalho, M., & Pinto, C. (2023). The Preservation of Memory and the Management of Information as a Step towards Sustainable Development. *WSEAS Transactions on Environment and Development*, *19*, 341–349. https://doi.org/10.37394/232015.2023.19.30

McBride, N. (2018). Is Information Systems a Science? *Communications of the Association for Information Systems*, *43*(1). https://doi.org/10.17705/1CAIS.04309

Montenegro, Z., Marques, J., & Sousa, C. (2024). *Integrating industrial tourism into active industrial companies: Advantages, barriers, and strategic approaches using the delphi method.* https://doi.org/10.24857/rgsa.v18n11-229

Mueller, S. T., Hoffman, R. R., Clancey, W., Emrey, A., & Klein, G. (2019). *Explanation in Human-AI Systems: A Literature Meta-Review, Synopsis of Key Ideas and Publications, and Bibliography for Explainable AI* (arXiv:1902.01876). arXiv. https://doi.org/10.48550/arXiv.1902.01876

Museums in a Digital Age. (sem data). Routledge & CRC Press. Obtido 23 de janeiro de 2025, de https://www.routledge.com/Museums-in-a-Digital-Age/Parry/p/book/9780415402620

O'Raghallaigh, P., McCarthy, S., & Adam, F. (2017). SOCIOMATERIALITY: AN OBJECT-INSPIRED PROPOSAL FOR IS SCHOLARS. *Research Papers*, 1348–1362.

Parry, R., Page, R., & Moseley, A. (Eds.). (2020). *Museum Thresholds: The Design and Media of Arrival.* Routledge. https://www.routledge.com/Museum-Thresholds-The-Design-and-Media-of-Arrival/Parry-Page-Moseley/p/book/9780367498986

Ribeiro, F., & Silva, A. M. da. (2002). *A Avaliação em Arquivística: Reformulação teórico-prática de uma operação metodológica.* https://repositorio-aberto.up.pt/handle/10216/80609

Salse-Rovira, M., Jornet-Benito, N., Guallar, J., Mateo-Bretos, M. P., & Silvestre-Canut, J. O. (2024). Universities, heritage, and non-museum institutions: A methodological proposal for sustainable documentation. *International Journal on Digital Libraries*, *25*(4), 603–622. https://doi.org/10.1007/s00799-023-00383-0

Shmueli, G. (2010). To Explain or to Predict? *Statistical Science*, *25*(3), 289–

310. https://doi.org/10.1214/10-STS330

Silva, A. M. da. (2013). Ciência da Informação e comportamento informacional Enquadramento epistemológico do estudo das necessidades de busca, seleção e uso. *PRISMA.COM*, *21*, Artigo 21.

Stalder, F. (2000). Pandora's Hope: Essays on the Reality of Science Studies. *The Information Society*, *16*(3), 245–247. https://doi.org/10.1080/01972240050133698

Vol. 9 No. 1 Suppl. (2024): JIIM Bibliographic Guide 2018-2023 | Journal of Integrated Information Management. (sem data). Obtido 23 de janeiro de 2025, de https://ejournals.epublishing.ekt.gr/index.php/jiim/issue/view/2111

Шелегина, О. Н. (2022). Музейный мир России: Актуализация регионального опыта. *Культурно-Антропологические Исследования*, *2*, 19–25.

2

Collaboration Between Archivists and Scientists at the American Museum of Natural History

Maya Naunton, American Museum of Natural History, United States

The archive at the Vertebrate Paleontology Department of the American Museum of Natural History has grown organically over the 130-year existence of the department. While there have been efforts to organize and arrange the papers in the past, most of the archive has not been processed.

Figure 1. Overall view of the Department of Vertebrate Paleontology archive

Figure 2. One of the unprocessed cabinets

An Institute of Museum and Library Studies grant received by the department in 2020 and extended in 2023 supports the effort to process the papers in the archive and make them accessible to researchers.

Our education as archivists teaches us how to organize information without reference to any specific subject. However, to be truly effective, it is essential for the archivist to learn as much as possible about the subject matter of the papers contained in the repository. When the archives deal with scientific material, that becomes so much more important.

My previous studies did not include paleontology and so when I started this job in early 2022, I had to rely on collaboration of the members of the scientific staff of the department. Without their help, I would not be able to understand what I was looking at and to describe it in a meaningful way.

The earlier attempts to organize the material in the archive, some of them dating to the 1980's and 1990's illustrate the importance of that point. The effort was led by a retired archivist from the Rockefeller University Archive, Ruth Sternberg, who brought her professional knowledge to the project (Figures 3a and 3b).

In her work, she relied on volunteers to organize and rehouse the material. According to some department staff members who were around at the time, there was not much interaction between the volunteers and

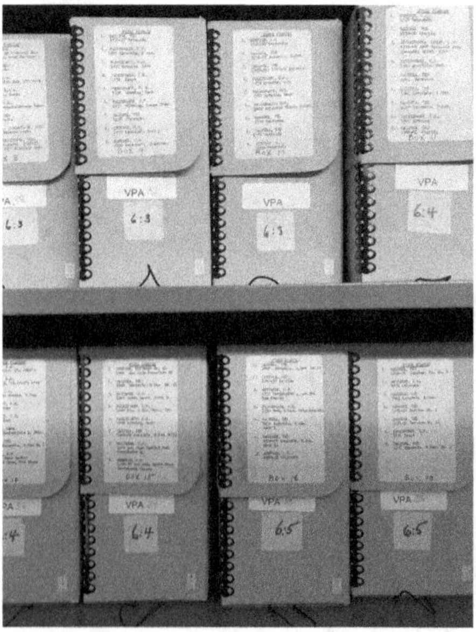

Figures 3a and 3b. Boxes with papers processed by Sternberg team

the science staff. This resulted sometimes in unfortunate, easily avoidable mistakes, as when a folder was labeled:

> **Subject**: Mammalia
> **Subset**: Dinosaurs

Other problems with the processed material were more subtle, but most of them resulted from the lack of communication between the people processing the archival papers and the staff of the department.

To ensure that I understood what I was looking at, I spent a lot of my time, especially in the beginning when we were doing the survey of the archive, talking to the staff of the department. Here I would like to acknowledge the help of Alana Gishlik, Carl Mehling, and Mick Ellison. They bore the brunt of my incessant curiosity, and they were endlessly patient and helpful in providing the information that I required.

Discussions with the science staff ranged from philosophical – what is an expedition? (Not an easy question despite seeming to be obvious and we will get back to that) - to specific questions – what is this a drawing of? What do I call it and why was it made? To answer that, Carl Mehling not only told me what animal was being illustrated – a Brontosaurus, he also found an old image of the galleries from the 1930's with that drawing in place (Figures 4 and 5).

But to go back to the question – What is an expedition? When I first started working in the archive, I was frustrated with lack of any historical information about the history of the department.

The work of the department largely consists of expeditions to collect fossils (Figure 6); work in the preparatory laboratory to extract those fossils from the stone matrix in which they were encased (Figure 7); exhibiting the fossils in the halls of the museum (Figure 8); and studying the fossils and publishing the results of the research (Figure 9).

Since expeditions form an important part of the history of the department, I felt that it was essential to reconstruct that information. The

Figure 4. An exhibition hall in 1939. /© AMNH

Figure 5. Pen drawing of the Brontosaurus. Artist and date unknown. Department of Vertebrate Paleontology Archive, VPA 107 Artwork. /© AMNH

Figure 6. Barnum Brown at Howe Quarry, 1934. 132806, AMNH Special Collections, Photographic Negative Collection/© AMNH

Figure 7. Staff preparing fossils. Early 20th century, Department of Vertebrate Paleontology Archive, VPA 104 Fossil Specimen Images, Series 1, Mounted Photographs /© AMNH

Figure 8. Short-legged Rhinoceros. 324822, AMNH Special Collections, Photographic Negative Collection /© AMNH

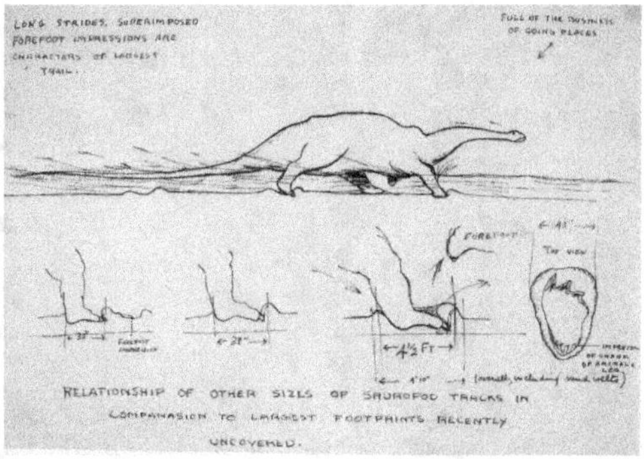

Figure 9. Diagram of dinosaur tracks by R.T. Bird. Ptc-6030, AMNH Special Collections, Photographic Transparency Collection/© AMNH

data has direct relevance to the ongoing study of the fossils, since it provides records of their collection and context. Not only that, but it is also important to the main archive of the museum, which holds some of the material relevant to the expeditions, especially photographic images. Describing those assets properly required consistent institution-wide metadata which was lacking.

After extensive discussions with the staff of the department, it was decided that what we would call an expedition for the purposes of the archive, would be a trip financed by the museum into the field whether that was for collecting fossils, reconnaissance, or to study stratigraphy (Figure 10). We excluded trips that were taken to study fossils in other museums, or to examine fossils that were offered to the museum, or to attend conferences.

However, while the question of which trip was to be called an expedition was resolved, other issues remained. When the staff recorded expedi-

Definition of Expedition

- Financed by the American Museum of Natural History
- Collecting fossils
- Surveying trips
- Trips to study stratigraphy

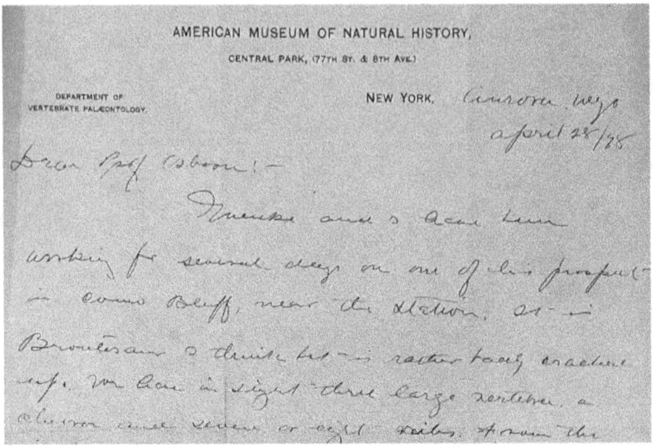

AMERICAN MUSEUM OF NATURAL HISTORY,
CENTRAL PARK, (77TH ST. & 8TH AVE.)
DEPARTMENT OF VERTEBRATE PALÆONTOLOGY.
NEW YORK, Aurora, Wyo.
April 28/98
Dear Prof Osborn:—
Menke and I have been working for several days on one of his prospects in Como Bluff, near the station. It is Brontosaur I think but is rather badly crushed up. We have in sight three large vertebrae a chevron and seven or eight ribs. From the

Figure 11. Letter from Granger to Osborn, 1898, Department of Vertebrate Paleontology Archive, VPA 101 Field Work, Series 2 Field Correspondence/© AMNH

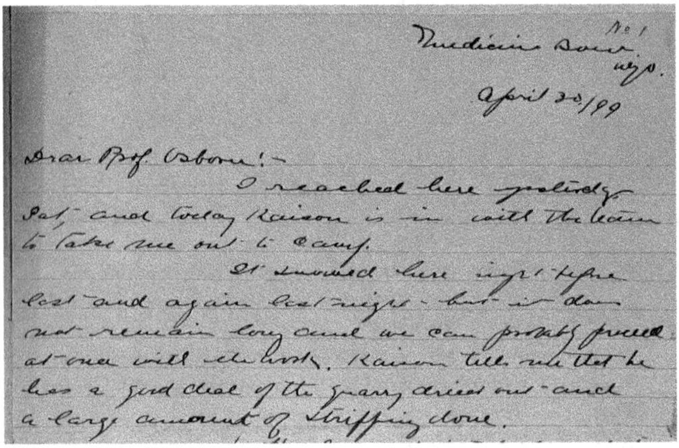

No 1
Medicine Bow
Wyo.
April 30/99
Dear Prof. Osborn:—
I reached here yesterday Sat. and today Kaison is in with the team to take me out to camp.
It snowed here night before last and again last night – but it does not remain long and we can probably proceed at once with the work. Kaison tells me that he has a good deal of the quarry dried out and a large amount of stripping done.

Figure 12. Letter from Granger to Osborn, 1899, Department of Vertebrate Paleontology Archive, VPA 101 Field Work, Series 2 Field Correspondence/© AMNH

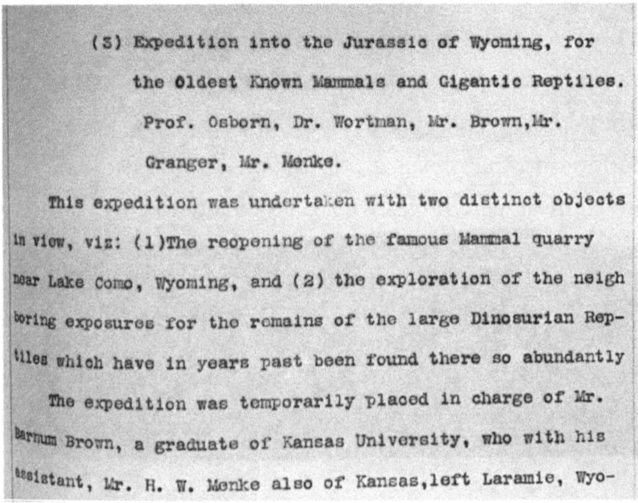

(3) Expedition into the Jurassic of Wyoming, for the Oldest Known Mammals and Gigantic Reptiles. Prof. Osborn, Dr. Wortman, Mr. Brown, Mr. Granger, Mr. Menke.

This expedition was undertaken with two distinct objects in view, viz: (1)The reopening of the famous Mammal quarry near Lake Como, Wyoming, and (2) the exploration of the neigh boring exposures for the remains of the large Dinosurian Reptiles which have in years past been found there so abundantly

The expedition was temporarily placed in charge of Mr. Barnum Brown, a graduate of Kansas University, who with his assistant, Mr. H. W. Menke also of Kansas,left Laramie, Wyo-

Figure 13. Departmental Annual Report 1897. Department of Vertebrate Paleontology Archive, VPA 102 Administrative and Personnel Records, Series 1 Annual Reports/© AMNH

SECOND EXPEDITION FOR DINOSAURS.

The sum of One thousand six hundred and sixty three dollars and twenty four cents was spent upon this expedition which was in the highest degree successful, under the able direction of Dr. Wortman assisted by Mr. Granger, Mr. Menke, Mr. Norman Grant who volunteered his services, Mr. Peter Kaison and Mr. Thompson the cook.

Figure 14. Departmental Annual Report 1898. Department of Vertebrate Paleontology Archive, VPA 102 Administrative and Personnel Records, Series 1 Annual Reports/© AMNH

tions whether it was in the annual reports, or in correspondence, they often referred to the same trip by different names. Sometimes they would mention a sponsor, sometimes it was called by the geological period that was being studied, yet at other times they referred to it by the geographical location.

As an example, I will discuss expeditions to a site that is famous in the Paleontological circles – namely the "Bone Cabin Quarry". The quarry is located northwest of Laramie, Wyoming and the department collected fossils there from 1897 to 1902. In the correspondence from the field, the participants identify the location where they are based: Aurora, Wyoming (Figure 11), in one case, Medicine Bow in another (Figure 12). In the departmental annual reports, the same expeditions are referred to as "Expedition into Wyoming" (Figure 13), Second expedition for Dinosaurs" (Figure 14), "Jurassic of Southern Wyoming expedition", and "the Bone Cabin Quarry" expedition. Deciding which expedition was being talked about in each document required a lot of help from my colleagues in the department. As the result of this work, which is still ongoing, the metadata librarian in the main museum library, Iris Lee and I are creating agent records for the expeditions that will eventually contain the names of the participants, dates, and locations and any alternative names that the expedition was referred to (Figure 15).

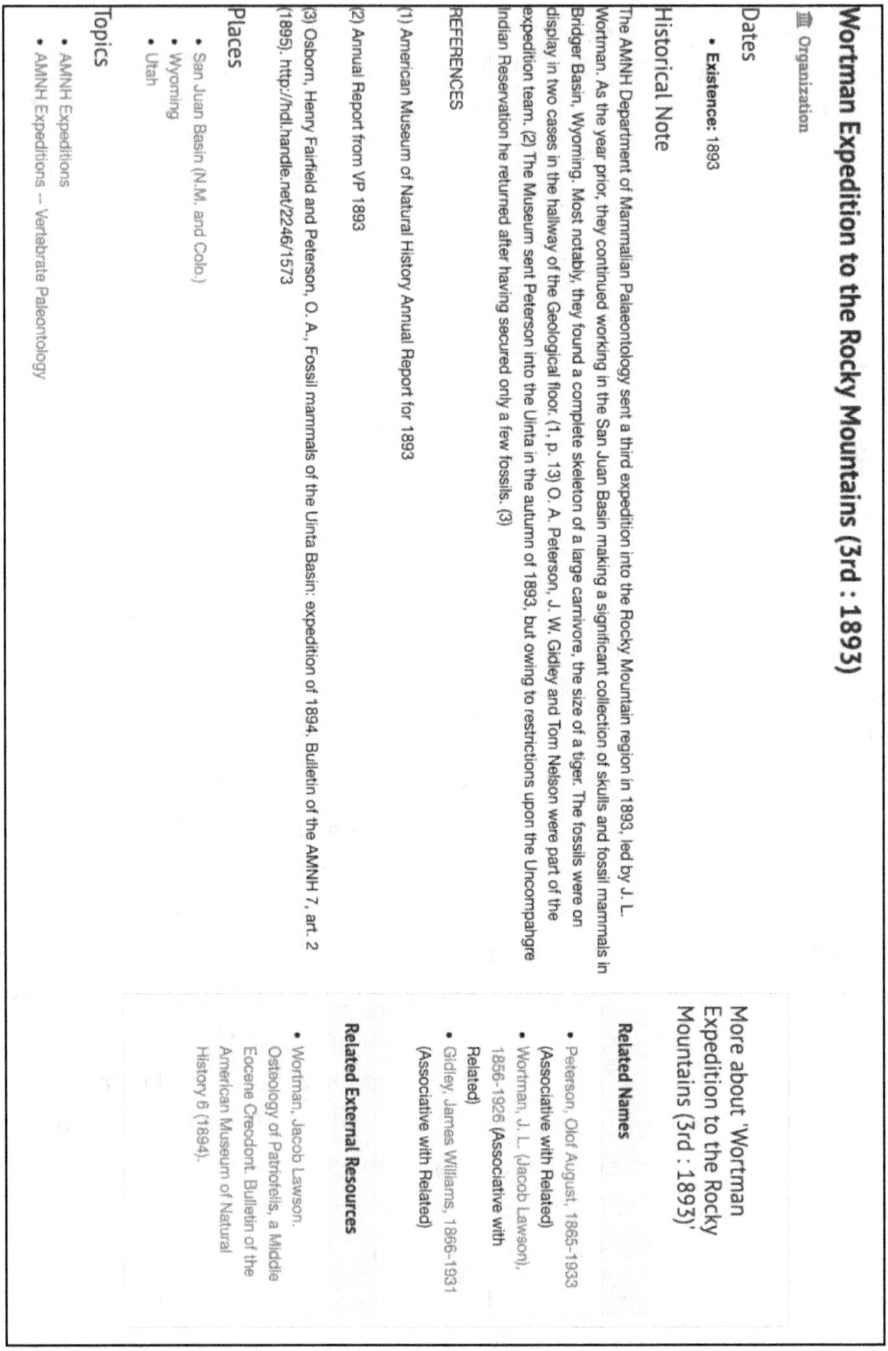

Wortman Expedition to the Rocky Mountains (3rd : 1893)

Organization

Dates

- **Existence:** 1893

Historical Note

The AMNH Department of Mammalian Palaeontology sent a third expedition into the Rocky Mountain region in 1893, led by J. L. Wortman. As the year prior, they continued working in the San Juan Basin making a significant collection of skulls and fossil mammals in Bridger Basin, Wyoming. Most notably, they found a complete skeleton of a large carnivore, the size of a tiger. The fossils were on display in two cases in the hallway of the Geological floor. (1, p. 13) O. A. Peterson, J. W. Gidley and Tom Nelson were part of the expedition team. (2) The Museum sent Peterson into the Uinta in the autumn of 1893, but owing to restrictions upon the Uncompangre Indian Reservation he returned after having secured only a few fossils. (3)

REFERENCES

(1) American Museum of Natural History Annual Report for 1893

(2) Annual Report from VP 1893

(3) Osborn, Henry Fairfield and Peterson, O. A., Fossil mammals of the Uinta Basin: expedition of 1894. Bulletin of the AMNH 7, art. 2 (1895). http://hdl.handle.net/2246/1573

Places

- San Juan Basin (N.M. and Colo.)
- Wyoming
- Utah

Topics

- AMNH Expeditions
- AMNH Expeditions -- Vertebrate Paleontology

More about 'Wortman Expedition to the Rocky Mountains (3rd : 1893)'

Related Names

- Peterson, Olof August, 1865-1933 (Associative with Related)
- Wortman, J. L. (Jacob Lawson), 1856-1926 (Associative with Related)
- Gidley, James Williams, 1866-1931 (Associative with Related)

Related External Resources

- Wortman, Jacob Lawson. Osteology of Patriofelis, a Middle Eocene Creodont. Bulletin of the American Museum of Natural History 6 (1894).

Figure 15. Screen shot of the ArchivesSpace record for one of the expeditions

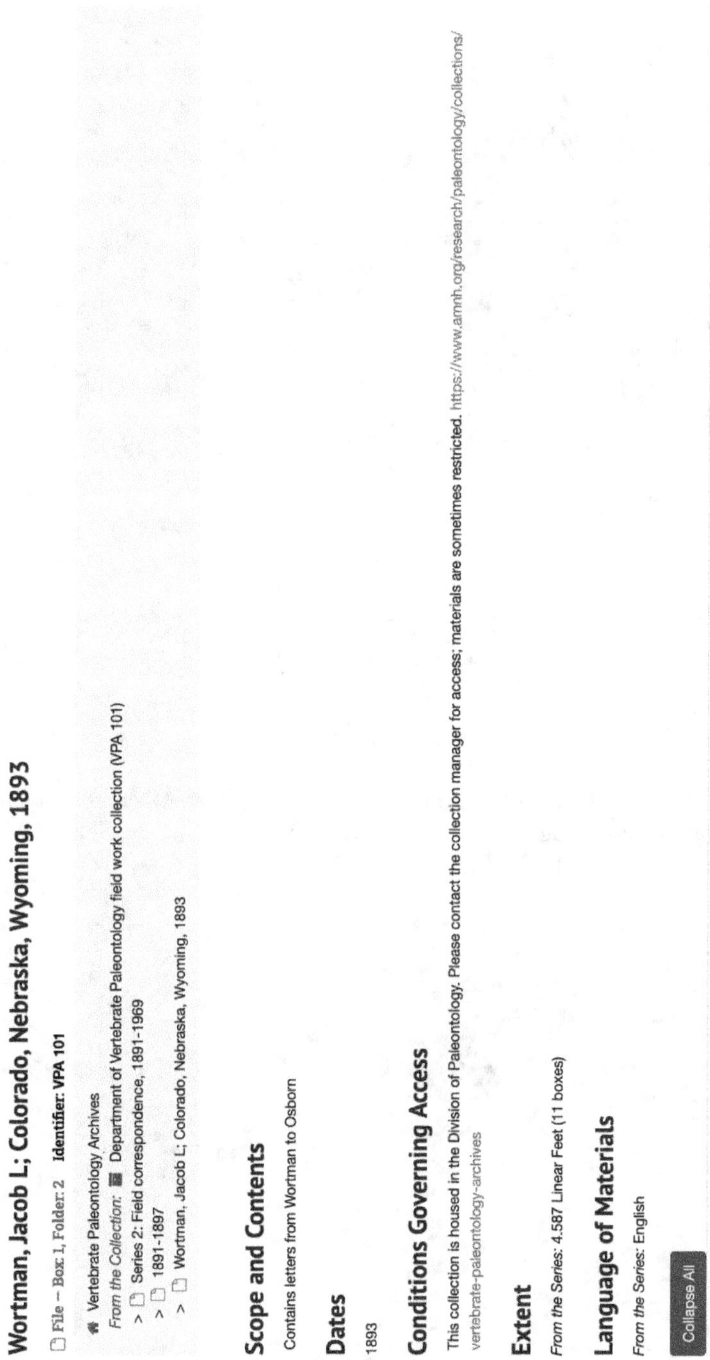

Figure 16. Screen shot of the ArchivesSpace record for field work correspondence linking to the expedition instance

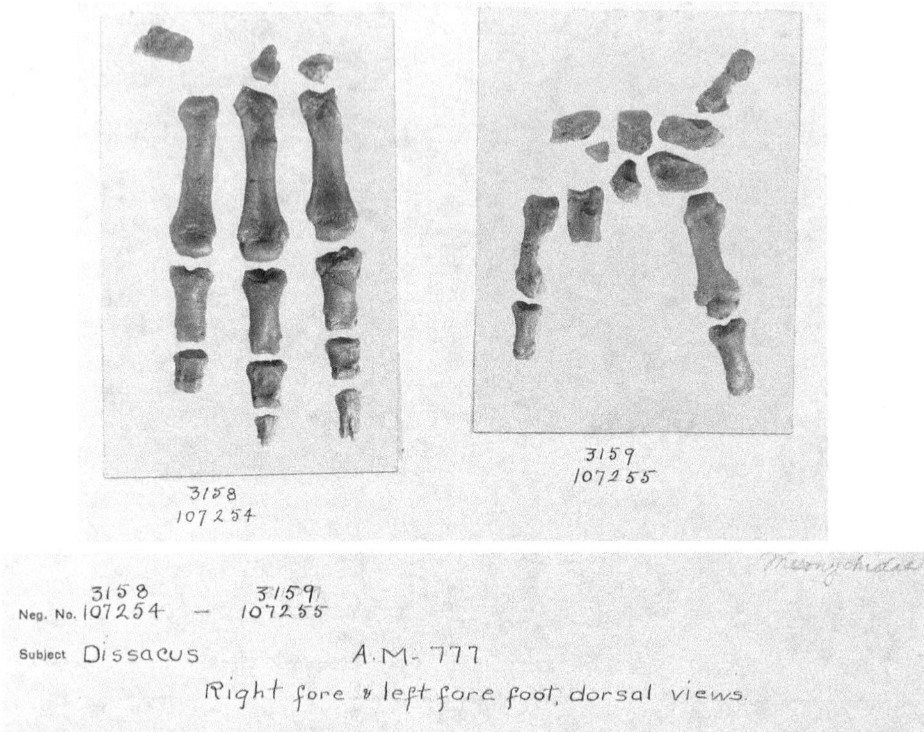

Figure 17. Front and back of the mounted photograph of a specimen. Department of Vertebrate Paleontology Archive, VPA 104 Specimen Images, Series1 Mounted Photographs/© AMNH

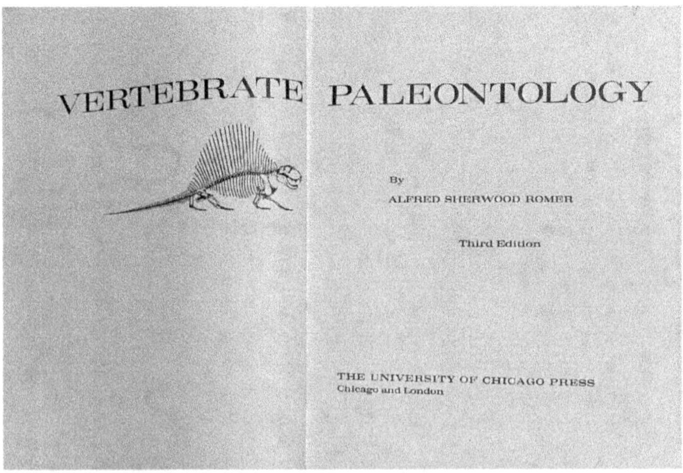

Figure 18. Title page, Romer, Alfred Sherwood, *Vertebrate Paleontology*, Chicago and London: University of Chicago Press, 1966, Third Edition

Figure 19. https://www.mensaforkids.org/teach/lesson-plans/classifying-animals/, accessed on 01/15/2025

As we process the papers in the Department's archive, we will be able to link to those expedition instances, as for example, in expedition correspondence, or field notebooks (Figure 16).

Another example where the help of my science colleagues was essential was processing a collection of images that were mounted on cardboard (Figure 17). We are still unsure when the images were gathered into this

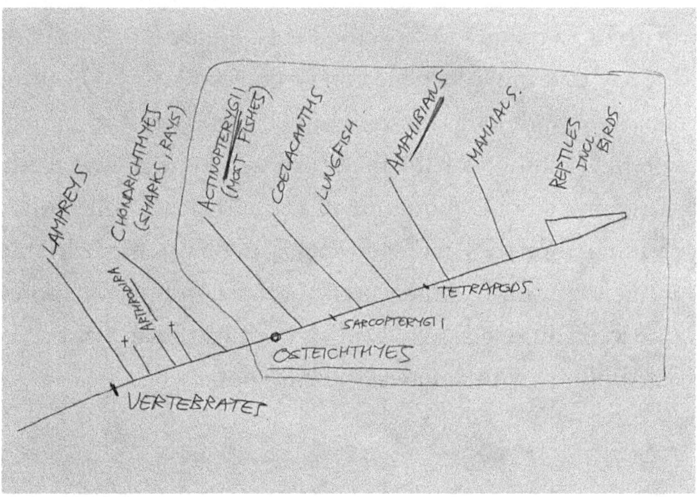

Figure 20. Drawing by Dr. Benson

collection, but correspondence from the department mentions people organizing them in the 1930s and 1940s. The department has an ongoing teaching relationship with Columbia University, and it is possible that they were used for graduate paleontology classes held in the museum.

The images were arranged according to Romer classification (Figure 18). Even though this classification has been superseded by more recent research, it was decided that the original order was to be preserved for historical reasons. Not much needed to be done with the collection physically, besides putting the images in new boxes, but issues came up when trying to describe the collection in ArchivesSpace.

I wanted to group the images so that the researchers would know where to look for a specific animal. The problem was that I have no idea on how the animals are grouped and the information on the physical cards with the images was not consistent. This required that I get a flash course on animal classification schemes. It was decided that the images were going to be grouped on the subclass level (Figure 19), which was all very good, but still left me not always knowing which subclass some animals were to be assigned to.

This in turn led to an impromptu lecture from our curator in charge of Fossil Amphibians, Reptiles and Birds Dr. Roger Benson and the following sketch to illustrate the lesson (Figure 20). This is the result of those ongoing consultations. The images are grouped in Class and then Sub-class, which will facilitate the process of searching for specific items (Figure 21).

I am happy to say that the information flow goes both ways. I share what I find in the archives through blogs on the museum's website and in personal conversations with the members of the department. In one instance, where a staff member is writing a book about one of the early fossil collectors, R.T. Bird, I was able to share sketches and correspondence from the field that was of help. As we continue to process and refine the organization of the material in the archive, I hope to be able to assist scientists both within and outside of the museum to use the bounty of the material that is contained in it.

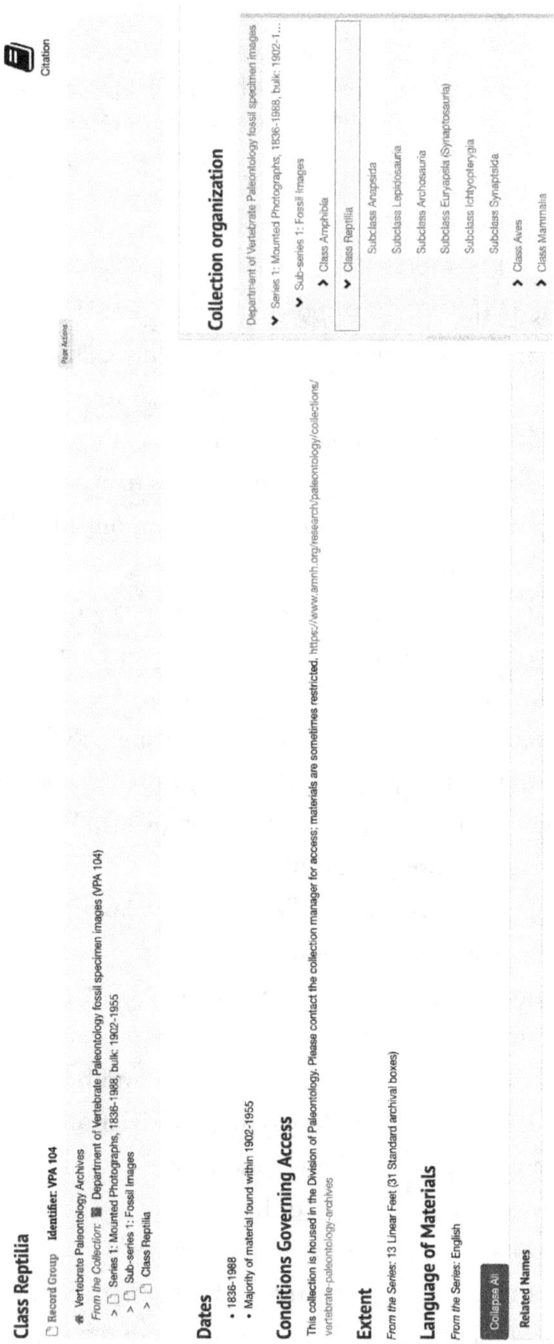

Figure 21. Screen shot of the ArchivesSpace record for the collection

3

Restrictions and Data in the Vertebrate Paleontology Archive

Kiana Clark, American Museum of Natural History

A key goal of the IMLS grant supporting our project is to improve access to the collections in the Vertebrate Paleontology Archive, chiefly by cataloging the records in the database, ArchivesSpace. While a sizable portion of the archive remains uncatalogued and not web-searchable, the creation of multiple finding aids and the presence of a dedicated archives team has led to an influx of reference requests from researchers from inside and outside of the museum. This broadened public audience was a marked success, but some concerns remained about protecting the legal interests of the museum, as well as the personal information and intellectual property represented in the materials of departmental staff. Protecting personal privacy is a core tenet of the archival profession's ethical obligations; while there is some legal codification around cases of copyright and intellectual property, many archivists balance access with privacy sensitivity by following a combination of institutional policy and theoretical frameworks proposed by authors like Laura Millar, Anne T. Gililand, and Judith A. Weiner. In the Vertebrate Paleontology Archive, the privacy concerns included not just typical personally identifiable information (PII) but also potentially sensitive information specific to the field of paleontology and the research represented in the archive.

After formalizing our access and use policies and bringing on a reference archivist, we began working on developing a restrictions policy that would act in tandem with our reference activities and processing projects. In order to create a comprehensive policy that suited the needs of the archives' users, as well as institutional policies, we followed the recommendations outlined in Anne T. Gilliland and Judith A. Weiner's "A Hidden Obligation: Stewarding Privacy Concerns in Archival Collections Using a Privacy Audit," which identifies legal, quasi-legal, and pragmatic privacy concerns for archives and outlines methods of locating and assessing them. We consulted a wide range of people throughout the Vertebrate Paleontology Department and the American Museum of Natural History. This included archivists from other departments, who have developed similar restriction policies; counselors from the Legal Department who could advise us on personnel and FERPA records; research associates, who are some of the most frequent users of the archive; and departmental curators, whose research is closely aligned to many of the materials in the archive.

For good measure, we also reached out to museums with similar archival collections to see how they governed restrictions and usage of sensitive materials, if they did so at all. While few other institutions were able to provide restriction policies, the core aspects of our policy came from other archivists in the AMNH. These included term restrictions on common record types like employee PII, financial records, meeting minutes from the Museum's Board of Trustees, and information protected by FERPA; these types of records are not specific to the field of vertebrate paleontology and thus allowed for greater cohesion with the archives of other departments, as there is often overlap between departmental collections. Our consultations with the Legal Department were particularly valuable for establishing clarity and context on institutional approaches to the broader legal mandates of FERPA and HIPPA; by examining medical and student records present in the archive, they helped us identify cases where we should deaccession or destroy materials. They also approved the use of separation sheets, a documentation of the restriction process that helps to maintain the integrity of the historical record (Gilliland and Wiener, 2014, 29). The Museum does not currently have an institution-wide records management program, so it was important for us to create cohesion among the departmental archives as a foundation for a program in the future. Seeking out perspectives from across the museum also allowed us to connect with the administrators who hope to develop the records management program and now know to incorporate archival perspectives and principles.

In addition to the standard restriction types, our conversations with curators in the Vertebrate Paleontology Department led to concerns about necessary restrictions related to the field of paleontology, most notably materials that involved fossil excavations and field work. Our collections include field notebooks, field correspondence, and research notes that often contain information about the localities, or sites, where excavation is taking place. There are particular concerns around making that information publicly available because fossils can be extremely valuable, and there have been many cases of looting and illegal collecting at both active and inactive dig sites. Active dig sites, which paleontologists from the AMNH may visit many times over the course of several years, are of particular concern as looting may cause damage to specimens that haven't been unearthed yet. One such example is the work being done in the Gobi Desert of Mongolia; AMNH paleontologists first visited the Gobi in the 1920s and have been returning annually since 1990, and our archive does contain field notebooks from these visits, and poaching has been a particular problem in the area.

While we aim to make the archive publicly accessible, it's important to balance the realities of paleontological research with archival use principles. To do this, we developed a set of guidelines to follow when determining

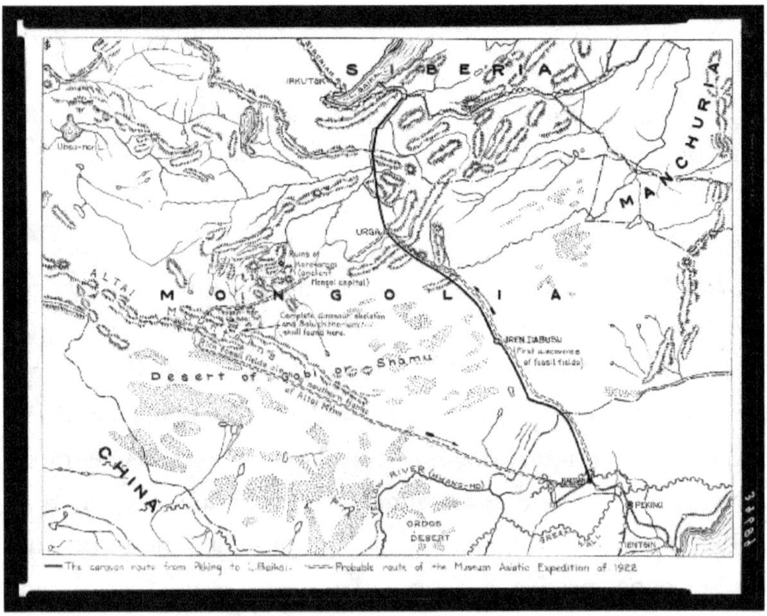

Figure 1. Map of probable expedition route for the First Central Asiatic Expedition to the Gobi Desert in Mongolia, 1922, Image no. 39689, AMNH Special Collections, Photographic Negative Collection /©AMNH

restrictions on materials with locality information. The very first step of this process is to determine whether the locality information is sensitive or not. This is done primarily by considering a set of criteria proposed by the Society of Vertebrate Paleontology. These criteria, officially titled "Best Practice Guidelines for Repositing and Disseminating Contextual Data Associated with Vertebrate Fossils," support the reproducibility of paleontological research by outlining cases in which locality data can be shared publicly. While the SVP encourages paleontologists to collect and archive extensive contextual data, they recognize that in some cases, public access to that data can cause harm to fossils, ongoing research, or non-paleontological resources like endangered species or ecosystems (Society of Vertebrate Paleontology, 2016). Two sets of criteria are provided to determine potential harm; the first is for site vulnerability, which indicates the potential of harm from human activity, while the second helps determine the sensitivity of data.

Many of the questions used for the criteria would require extensive research for a lay archivist, particularly around the uniqueness and scientific importance of specimens or whether sites are active or inactive. In addition to

A. Criteria for Site Vulnerability

An answer of 'yes' to any of the following criteria indicates potential of harm from human activity. The more 'yes' answers, the more likely the impact of harm there is.

1. Are there taxa that occur only at this site?
2. Do rare taxa occur at this site?
3. Does the site yield unusually complete or exceptionally well preserved fossils?
4. Is this site a unique or rare exemplar of its geological age or geographic provenance?
5. Is the site likely to produce additional specimens that will add significantly to the knowledge of variation within a taxon or to the completeness of a paleocommunity?
6. Are contextual field data that remain at the site easily damaged or destroyed?
7. Is the excavation or contextual documentation of the site still in progress?
8. Are scientifically important fossils that remain at the site easily damaged?
9. Is the site easily located because of proximity to roads or to easily visited or identifiable landmarks?
10. Are there non-paleontological resource values that would be impacted by releasing the location of the site?

Figure 2. Excerpt of the criteria for determining site vulnerability according to the Society of Vertebrate Paleontology, "Best Practice Guidelines for Repositing and Disseminating Contextual Data Associated with Vertebrate Fossils," Bethesda, 2019

consulting the experts in our department, we have developed other strategies and resources to determine the sensitivity of locality information in our archive. Based on conversations with the Curator-in-Charge of the archive, our considerations focus primarily on the rarity of specimens, the current status of the locality, and previous publication of data. Online resources such as Mindat.org, hosted by the Hudson Institute of Mineralogy, and the Paleobiology Database, a multi-institutional project, have been valuable for providing data on specimen instances, meaning that we can search a particular specimen and see where and when it has been found. If the only instance of a specimen was found in a locality described in our archives, it indicates that it may be a rare taxon, making the site more vulnerable. The current status of localities, in terms of their ongoing importance and use in research, is typically determined by consulting with members of our department who conduct field work - any archival materials containing information about the sites they annually visit would be restricted. These conversations also help to establish respect for the intellectual property created by staff members for the museum and support the development of further research and scholarship (Millar, 2017, 108). When sharing their informal policies around sharing locality data, Patricia A. Holroyd, the Senior Museum Scientist at the University of California Berkeley Museum of Paleontology, noted that many once-historical dig sites in their local area have been developed - searching exact coordinates on Google Maps is an easy way to rule out restrictions or determine how accessible the site would be. Finally, we search for journal articles that reference the locality site or specimens to see if coordinates or maps have already been published; if our archival materials in conjunction with published information create a more detailed picture of the locality, the materials will likely be restricted in order to avoid enabling access to the site.

When one or more of these factors indicate that the site is vulnerable and sensitive, the materials become restricted, meaning we place the physical materials in a flagged shadow folder that can easily be removed from a box requested by a visitor. Visitors can still seek approval to view restricted materials from the Curator-in-Charge of the archives, who is a paleontologist by trade; because we require visitors to provide details on their academic credentials and research plans, he can make a higher assessment of the request and the context of the restricted materials to determine what kind of risk might be associated with sharing. If researchers are approved to

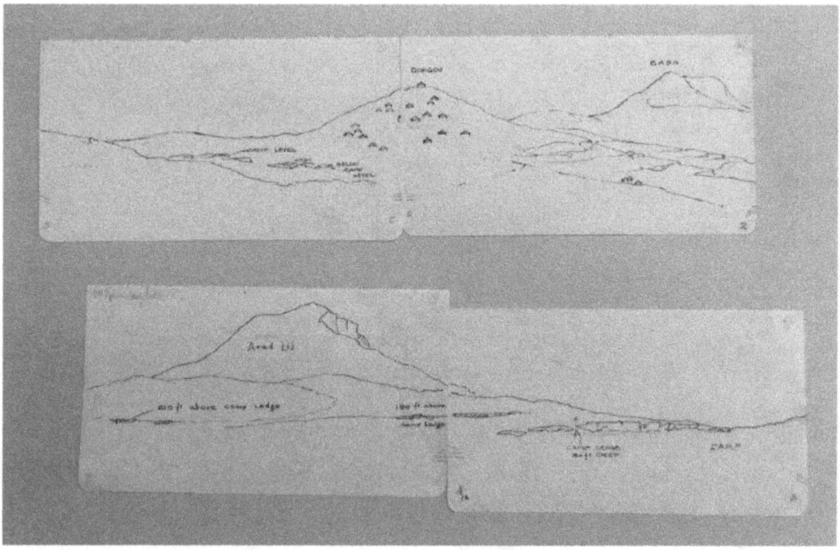

Figure 3. Supporting illustration of field camp from Barnum Brown's field notes from the Dudley Expedition to Ethiopia in 1921, Department of Vertebrate Paleontology Archive, VPA 114 Barnum Brown Papers, Series 3 Field Work/ ©AMNH

view restricted materials, they are prohibited from taking any photographs to further limit the risk of disseminating sensitive locality information.

While access to our archival materials is primarily facilitated by the publication of finding aids on ArchivesSpace, we hope to increase accessibility in the future by creating a workflow for the museum's newly funded digital asset management system (DAMS). Similarly to our restrictions project, this is a multi-department project involving IT, digital archivists from the main library, and data managers for the scientific collections, with the ultimate goal of creating a centralized location for the different kinds of collection data held across the institution. This includes scientific information like specimen data, digital objects including 3D scans of specimens, and archival

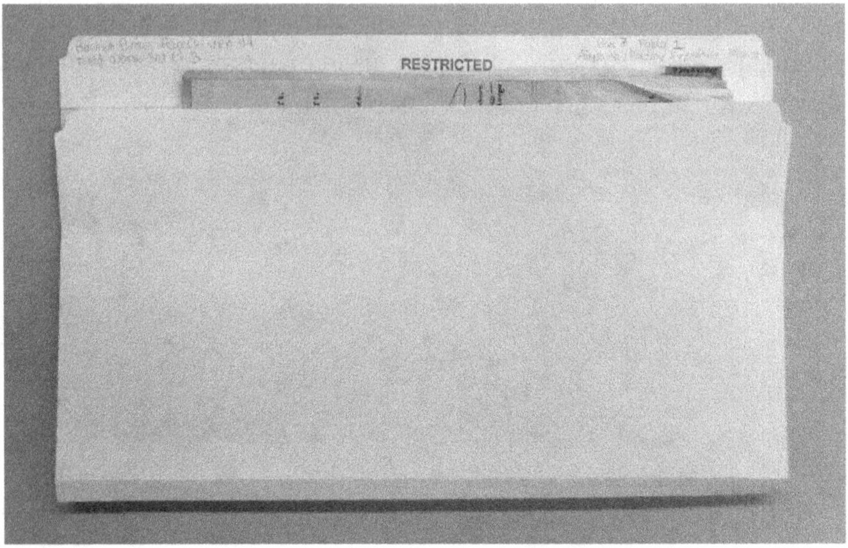

Figure 4. Restricted folder containing field materials and sensitive locality information from the Barnum Brown Papers in the Vertebrate Paleontology Archive (Courtesy of Kiana Clark)

materials that provide historical context to the scientific collections. This is a long-term project with many obstacles; the institution's collections are siloed across departments, there is a huge range of digital and physical formats, and various databases that are incompatible with each other are already in use, but we are dedicated to telling a more comprehensive story about the collections that incorporates all the relevant digital, historical, and scientific elements.

Bibliography

Anne T. Gilliland and Judith A. Wiener, "A Hidden Obligation: Stewarding Privacy Concerns in Archival Collections Using a Privacy Audit," *Journal for the Society of North Carolina Archivists* 11, no. 1 (Spring 2014): 19–35.

Millar, Laura A. "Balancing Access and Privacy." In *Archives: Practices and Principles*. (Cambridge: Facet Publishing, 2017).

Society of Vertebrate Paleontology, "Best Practice Guidelines for Repositing and Disseminating Contextual Data Associated with Vertebrate Fossils" (Bethesda, MD, December 2019).

4

Beyond the Walls and Changing the Space of Archive Through the Lens of The Adler Museum of Medicine in South Africa

Samuel Umoh Uwem, University of Hradec Kralove, Czech Republic
Adetola Elizabeth Umoh, University of Autonomous Barcelona, Spain

Abstract

Drawing from the Adler Museum of Medicine in South Africa, this paper briefly provides an overview of the Center's, functions and policies that governs the archives housed at the museum. The paper examines how archival practice through digitization strengthens or undermines provenance and original order. It analyses the migration to digitization of objects, the procedures, conditions, and reasons that influence the choice to include or exclude medical archives, and how they incorporate these insights into their digital archives' design, functionality, and affordance. Based on oral history, archival sources, documentaries, and interviews in South Africa, this paper argues that since COVID-19, the medical museum archives are simultaneously trying to digitize their holdings and maintain the physical originals. However, there are ethical concerns on digitalization and concerns amongst Archives and Records Management (ARM) practitioners about a coming digital dark age: a phenomenon where a significant amount of crucial archival information is lost due to both rapid digitization of information and lack of reliable long-term preservation methods.

Introduction

Many Galleries, libraries, archives, and museums (GLAMs) globally are digitizing their collections to protect and conserve the objects while simultaneously making them accessible to visitors in a form that is immersive and captivating for the usage and reuse of digital content (Keene and Parle 2015; la Porte and Higgs 2019, Mabe et al. 2021). Coupled with reducing the financial burden associated with digitization (Keene and Parle 2015, Mabe et al. 2021). Digitalization entails converting analog or manually produced materials, including documents or photos, into digital format for electronic access (Nkholedzeni 2019).

Medical museums are organizations that house and display artifacts connected to medicine or health that are of historical, scientific, artistic, or cultural significance (Nkholedzeni 2019, Tunç 2023). Universities with medical museum archives have amassed and continue to amass many artifacts associated with their research and teaching endeavors. In addition to medical photographs and the technologies used to capture them (such as X-ray equipment), displays frequently feature models, instruments, books, and manuscripts (Tunç 2023). South African medical museums are not immune from the digitalization trend as a means of preservation and improving access to their collections in South Africa (la Porte and Higgs 2019, Mabe et al. 2021). COVID-19 also triggered the need for digitalization (Momoti and Marutha 2022). Despite the perceived benefit of digitalization, preservation of context is a challenge in the digital depiction of medical archive sources. Two essential archival arrangement principles that support the preservation of context in collections are original order and provenance. Provenance explains the origins of archive records, whereas original order enables the illumination of linkages between records (Douglas 2018; Ghaddar 2022). However, many archives have found it challenging to digitally reproduce the tactile pleasure of looking through an archive. This difficulty has occasionally led to context loss or the inability to identify the initial order digitally.

Overview of Adler Museum of Medicine

The Adler Museum of Medicine, founded in 1962, is in the South African Institute for Medical Research, Johannesburg. The Adler Museum of Medicine was formally turned over to the University of the Witwatersrand in June 1974. The Museum bears the name of Dr. Cyril Adler, a well-known physician and medical historian from Johannesburg (Keene 2009). The Adler Museum of Medicine chronicles the history of the health sciences in Southern Africa. The Museum is a component of the University of the Witwatersrand's (Wits) Faculty of Health Sciences in Johannesburg and has an important educational role (Keene 2009). It also serves as a vital instructional resource for university students and other levels of education. Through collections, research, teaching, exhibitions, and publishing, the Museum enhances the University's educational offerings, particularly in the Health Sciences. The museum's displays provide a distinctive perspective on the development of South African medical practices and the function of medicine in society. The museum also teaches visitors about the history of medicine and the development of medical practices in South Africa (Keene 2009).

The Museum is an excellent resource for teachers because of the diverse information and sources pertinent to the Revised National Curriculum Statement. In addition to being interesting from a historical perspective, the hundreds of fascinating exhibits on display showcase technological advancements in the medical and healthcare industries. A key component of the curriculum is the use of evidence, and students can study authentic or replica sources and learn how to analyze evidence by visiting the Museum.

This also alludes to South African schools' curricula modification, and the country's adoption of outcome-based education led to an increasing number of teachers establishing connections with museums in some subject regions after the first democratic election in 1994 (Keene 2009).

The Museum also addresses some topics covered in South African school curricula: significant discoveries, technical advancements, indigenous knowledge, and traditional healing (Keene 2009).

The collection has 40,000 artifacts that illustrate the development of pharmacy, dentistry, ophthalmology, and medicine over time (Keene 2009). It enhances the museum's research and educational offerings, particularly those of the Faculty of Health Sciences, through collections, research, instruction, exhibitions, and publications.

Collection of Adler Museum of Medicine

The Adler's Museum's holdings comprise three categories: the study collection, the permanent collection, and the library – archives (Keene 2009). The permanent collection comprises microscopes and other scientific instruments, early bleeding and cupping tools, including a finely crafted incision knife, 17th-century ceramic pharmacy jars, an exquisite collection of bone China and ceramic feeding cups from the 18th and 19th centuries, and other historical items are on display. The Adler Museum has valuable and fascinating collections that show the evolution of pharmacy, dentistry, optometry, and medicine across time. Scientific instruments, microscopes, early bleeding and cupping equipment with an exquisitely crafted incision knife, 17th-century ceramic pharmacy jars, a collection of bone China and ceramic feeding cups, some from the 18th and 19th centuries, a wooden-handled amputation set from the early 19th century in a wooden case, diagnostic and surgical tools, treatment devices like one marketed as a "Patent magnetic electrical machine for nervous diseases" that Queen Victoria used to relieve her rheumatism in the 19th century, and the first electrocardiograph machine (1917) are among the medical historical items on display. It contains early anesthetic equipment, brass ear syringes and ear trumpets (early 20th century), hospital and nursing equipment, medical ephemera, the first artificial kidney machine in South Africa, and the Johannesburg General Hospital.

Study collection

This includes objects and other things that are either superfluous duplicates of items in the permanent collection or are in relatively bad condition. This collection is utilized for demonstrations and educational purposes with audiences of all ages.

The Archives and Library

The Museum features a rare book collection and a reference library on medical history comprising about 5,000 volumes and journals on the subject. The archives include records of the museum's past, as well as a comprehensive subject archive that includes images and biographical data on medical and other health care. It also contains records from the closed Transvaal Memorial Hospital for Children. The library contains rare volumes, and a sizable history of the medicine reference section can be found in the Museum library. Additionally, the library has an archive that provides students, academics, and the public with biographical details of thousands of medical and allied health professionals. The Museum features an important health sciences reference library history and a collection of rare books. The library has a subject-matter-organized archive. Photographs, notes, academic credentials, documents, personal papers, and memorabilia of notable academics and health professionals are among the thousands of biographical materials about medical and allied health professions that are available for research. The Museum offers first-rate facilities for teaching and researching the history of health sciences, as well as public lectures, tours, and temporary exhibits. The Museum augments the educational activities of the University, especially the Health Sciences, with its collections, research, teaching, exhibitions, and publications.

Digitization of Medical Museum Archive: Sine qua non

Digitalization by medical museums is increasing because of advancements in technology, making their collections widely accessible to visitors and COVID-19. Digitization has an impact on improving the educational value of museum content and the memory-making element of museums (Meehan 2022). Digitization made accessing museum information easier for the general public (Meehan 2022, Adams and Kopelman 2022).

Digitalization altered the relationship between museums and their patrons, making this "relationship" more complex. In addition, creative applications of digitalization have raised museum visitors' expectations (Palumbo 2022, Adams and Kopelman 2022). Even while digitalization is widely applauded and seen as an advancement for museums and medical archive institutions, it must not be disregarded due to the possible risks. Unprecedented access is undoubtedly a positive development for museums, but it may also result in privacy violations concerning sensitive collections

Nonetheless, digitization is also impacted by ethical considerations such as the right to access information, privacy anonymity, and, recently, the South Africa POPI Act (Protection of Personal Information Act). As regards privacy, Keene and Parle (2015) note that many types of patient records, some of which were deleted by physicians from Wits teaching hospitals and clinics in the 1980s, are housed in the Adler Museum because the security police during apartheid had occasionally confiscated patient records, which were subsequently used to determine the patients' names, the types of injuries or diseases they had been diagnosed with, and the course of treatment they had received. The state security forces might determine whether the patients were political activists if they had access to these records. Patients with rubber bullet gunshot wounds, for instance, were probably involved in boycotts or protests where police opened fire on demonstrators. Additionally, the medical records of well-known captives are kept in two files at the Adler Museum.

Another consideration regarding digitalization relates to balancing access and anonymity and the urgent need to safeguard sensitive, significant medical records. This alludes to the consideration that digitization could also breach ethics, such as anonymity. The Adler Museum, for example, has numerous other records that include historical health-related information. For example, the registers of Chinese miners who were sent back to their home country for medical reasons by the Witwatersrand Native Labour Association, which is the recruiting arm of the South African Chamber of Mines. These registers are fascinating to read and provide a depressing account of life in mines, including the incidence of diseases, typical causes of accidents and injuries, and more. The problems facing contemporary museums and archives are much more extensive than those depicted by the aforementioned cases.

Storage space is the other area that needs attention. The National Institute of Occupational Health disposed of hundreds of X-rays during the 2004–2005 building clearance process for renovations, which found its way to the Adler Museum. The Adler Museum would not have been able to hold more than a small portion of the enormous collection due to the skip-loads of X-rays that were being dumped. Substantial historical (and possibly scientific and medically significant) material may be lost in the process, even if the need to relocate and repair it and the severe limitations on storage capacity are well acknowledged.

Policies have been put in place to safeguard medical archives, such as copyright bills as evidenced in South Africa's GLAMs eventually inclusion in South Africa's contested 2022 proposed copyright bill. The current copyright legislation in South Africa originated in 1978 and did not sufficiently address GLAMs. The 2022 copyright bill seeks to support authors' financial interests and keep up with technological developments. The Bill has suggested several amendments to the Copyright Act. Many versions of the Bill have been passed over the years, and opinions on whether the current version of the Bill is beneficial or detrimental to South Africa are divided.

Also the national strategy on the "Digitization of Heritage Resources" open for public consultation since August 2010 (DAC 2010). Since no national policy exists, no uniform local guidance guarantees proper standards and methods for digitizing content. A well-planned digitization strategy is needed for an institution to deliver a valuable and sustainable contribution to the digitization of cultural heritage. Good digitization management should include guidelines for digitization, collection management, access, metadata, intellectual property (including copyright and privacy issues), and how digitization and preservation should be performed as digitization policies recommend (la Porte and Higgs 2019).

However, there is skepticism about how digitalization accurately reflects the actual physical archival material. Due to this, medical museums may decide, for instance, to only scan the front of photos for online consumption. A comments field is then used to enter any annotations on the back of the picture. Although the notation is not shown in its original format, it is still maintained. Archivists could also inscribe short forms, handwriting interpretations, and transcriptions to record the notation as metadata. Human error and interpretation can affect the transcription process itself. This also alludes to concerns amongst Archives and Records Management (ARM) practitioners about the digital dark age: a phenomenon where a significant amount of crucial archival heritage, such as traditional medicine and knowledge, would be lost due to the digitization of information and lack of reliable long-term preservation methods.

Archives and Records Management (ARM) professionals are worried about the impending digital dark age, in which a large portion of cultural legacy is destroyed because of the quick digitization of information and the absence of trustworthy long-term preservation techniques.

An archival website with transcription guidelines may help allay concerns about inaccurate transcription or omissions. For cases of archival sources that are in machine-readable text, using OCR (Optical character recognition) software can eradicate some transcription errors and allow the documents to be full-text searchable, which can be invaluable for researchers. Nevertheless, OCR is not flawless, and mistakes are usually most noticeable when applied to older texts with crooked or faded typefaces. Many organizations lack the time or resources to correct OCR problems, which can take a great deal of time for archival staff.

It is argued that when certain portions of a collection have been digitized, archives should make it apparent that only a portion of the content is accessible online and provide information on how to access the remaining content. Many archives and heritage institutions are attempting to incorporate the digitization of conventional archival material into their daily operations. Developing successful and beneficial digitization projects requires understanding patron preferences and how they use already digitized content.

Other best practice guidelines and standards for digitization guide museum digitalization, including Dublin Core and Taxonomy and Naming for metadata (de la Porte and Higgs · 2019). Technological standards exist for digital video, music, 3D, and image formats (TIFF, JPEG, and GIF). Adler Museum also follows the guidelines of Wellcome Library Archives (London), a renowned primary resource on medical history, which has a valuable example policy on access to research data that describes the kinds of sensitive material that are kept in its archives (Keene 2009). South Africa's museum framework also employs a comprehensive set of UK principles. Additionally, once finalized in August 2014, this policy specifies the terms under which this information is made available and accessible on their website (Keene 2009). Ethics must be considered if digitization aims to increase access for a wider audience. Museums acquire sensitive collections containing vulnerable topics and individuals for many reasons; one may be to address complex histories in an ever-continuing pursuit of educating visitors. It may be argued that it is important for museums to obtain such collections. While on exhibition and in storage, museums can ensure, or strive to attempt, that sensitive collections and people affected can retain a sense of privacy and protection due to exhibition design or collaboration with source communities regarding storage conditions. However, through digitization, the ability to ensure privacy can be hindered. POPI (Protection of Personal Information Act) has detrimental effects on all archive organizations that gather, maintain, and use personal data for historical, statistical, or research purposes, despite its notable promotion of personal information security. President Jacob Zuma signed the Protection of Personal Information Act (POPI) into law on November 27, 2013. Since its initial drafting in 2004, POPI has experienced numerous revisions. The National Assembly and the National Council of Provinces have been exchanging the then-bill since it was initially introduced in Parliament in 2009 and in August 2013.

References

Adams, Tacy and Kopelman, Sarah. 2022. Remembering COVID-19: Memory, crisis, and social media. *Media, Culture & Society*, *44*(2), 266–285, https://doi.org/10.1177/01634437211048377

de la Porte, Beatrice, and Higgs, Richard. (2019) Challenges in the digitization of cultural heritage material in the Western Cape, South Africa. *South African Journal of Information Management*, *21*(1), 1-11. https://doi.org/10.4102/sajim.v21i1.1104

Douglas, J. 2018. "A call to rethink archival creation: exploring types of creation in personal archives," *Arch Sci* 18:29–49.

Ghaddar, J.J. 2022. "Provenance in place: crafting the Vienna Convention for global decolonization and archival repatriation." In Lowry, J. (ed), *Disputed Archival Heritage*. Obingdon: Routledge, pp 49–86.

Keene, Rochelle. 2009. "Educating students in a university museum environment: The Adler Museum of Medicine, Faculty of Health Sciences, University of the Witwatersrand, Johannesburg," *Med Secoli.* 2009;21(1):255-75. PMID: 20481369.

Keene, Rochelle and Parle, Julie. 2015. "Museums, archives and medical material in South Africa: Some ethical considerations," *South African Museums Association Bulletin* 37: 6-16.

Mabe, Kagiso and Andrea Potgieter. 2021. "Collaboration between libraries, archives and museums in South Africa," *South African Journal of*

Information Management 23(1), 1-8. https://doi.org/10.4102/sajim.v23i1.1269

Meehan, Nicole, 2022. "Digital museum objects and memory: Postdigital materiality, aura and value," *Curator: The Museum Journal* 65(2), 417–434. https://doi.org/10.1111/cura.12361

Momoti, Nikiwe, and Solomon Ngoako Marutha. 2022. "Whoever Controls the Past, Controls the Future: Archives and Records Management Practitioners' Response to COVID-19 Pandemic in South Africa," *Mousaion: South African Journal of Information Studies* 39 (4):18 pages. https://doi.org/10.25159/2663-659X/9964.

National Archives and Records Service of South Africa Regulations. 2002. Available at: https://www.nationalarchives.gov.za/node/6548861

Nkholedzeni Sidney Netshakhuma. 2019. "The role of archives and records management legislation after colonialism in Africa: Case of Southern Africa," *Records Management Journal* 29(2). https://doi.org/10.1108/RMJ-09-2018-0024

Palumbo, Rocco. 2022. "Enhancing museums' attractiveness through digitization: An investigation of Italian medium and large-sized museums and cultural institutions," *International Journal of Tourism Research* 24(2), 202–215. https://doi.org/10.1002/jtr.2494

Tunç Uğurgül. 2023. "Museums of Medicine and Health: Curating Public Health," *Infect Dis Clin Microbiol* 2023;1:69-81. 10.36519/idcm.2023.195

5

Curating the Clinical: Visual and Material Methods for a Photographic Collection

Michaela Clark, Department of the History, Philosophy and Ethics of Medicine | Institut für Geschichte, Theorie und Ethik der Medizin, Heinrich Heine University, Düsseldorf, Germany

Abstract

This chapter provides a conceptual and methodological overview of how an assembly of decontextualised clinical photographs held at a medical school in Cape Town (South Africa) was 'curated' to restore its research value. No records regarding the making, use, or disuse of this photographic collection remained. Thus, at the start of the project, it was recognised as 'orphaned' and at risk of abandonment. The collection's institutional custodians pushed back against this interpretation by putting forward that a merging of practical and intellectual handling with historical analysis (a process referred to here as curation) could reinstate its use-value. The clinical photographs are framed as both visual and material records or 'photo-objects' that, through a merging of archival processing and object-centred research, were recontextualized and so rendered a productive source for historical investigation. Ultimately, the chapter demonstrates that curation of this kind can serve as a primary step in researching orphaned photographic materials, thus offering a new approach to writing histories of visual and material medical culture in postcolonial contexts.

Introduction

This chapter describes the process and product of 'curating'[1] an assembly of historical clinical photographs collectively known as the University of Cape Town's Department of Surgery Collection of 20th Century Clinical Photographs (the UCT DoS Collection). The photographs are dated between 1901 and 1967 and were actively employed by the surgical division of Cape Town's medical school from the launch of clinical education at this facility and in the country as a whole (in 1920). Having fallen into disuse during the 1980s, the UCT DoS Collection receded from institutional awareness. Then, thirty years later, it was accidentally uncovered in a small storage room in the Department of Surgery by members of the university's Pathology Learning Centre (PLC).

The photographic records depict the heads, limbs, torsos, disembodied organs, as well as x-rays of patients. All are imaged in black and white, mounted on rectangular cardboard backings, and many are annotated with the details of the patient (their name, age, and disease) and the date the photograph was taken. Produced across multiple clinical teaching facilities, the collection showcases patients spanning age, class, race, gender, and diagnostic divides. After decades of disuse, however, the context of the collection's origins and purpose had been lost.

This chapter reports on the process involved in and initial findings produced by using archival processing methods in conjunction with provenance research to 'curate' the UCT DoS Collection. It offers a primary analysis of the visual and material evidence gleaned in this process by relating contextual hints to the historical actors, events, technologies, and practices involved in the making, use, and distribution of its contents. It sees photographs as entities that operate within an 'ecosystem' of production, use, storage, and exchange.[2] They thus possess a 'cultural biography'[3] and have 'social lives'[4] that may be traced.

While records of this kind are traditionally treated by archivists in a similar way to text-based records, this chapter frames them as 'photo-objects' – as complex items comprising the photograph "together with its mount and all its annotations and traces as a material object [...] that exists in space and time, and in social and cultural contexts."[5] It thus follows Elizabeth Edwards' advocacy that photographs can serve as a methodological and conceptual starting point for historical research: in her words, they are things to "think with."[6] Ultimately, the chapter maps this approach in relation to the UCT DoS Collection by demonstrating how curation has rendered useful a forgotten clinical photographic collection, thus offering a solution to the research-based problems associated with orphan works in Cape Town.

Making an Archive

Many historical collections accidentally discovered within institutional walls are turned away as archives are forced to contend with restricted space, diminishing budgets, and staff already stretched thin.[7] The cost of conservation materials as well as the extensive investment needed to assess, arrange, and describe new collections is often deemed to be better spent on existing holdings.[8] The result is that at-risk material may be left to lie disorganised, uncatalogued, and in various stages of decay – its contents unknown, its research potential untapped, or its very existence threatened.

The UCT DoS Collection was a victim of conditions of this kind. It had been accepted by the PLC between 2014 and 2019 after roughly 50 years of neglect. During this time, it was exposed to environments not conducive to preserving either text-based or photographic materials.[9] The collection's contents showed significant signs of decay, including silvering (degradation of photographic emulsion), exposure to damp, dust, and UV light (warping of paper, large watermarks, and areas of significant discolouration), as well as evidence of damage caused by pests (emulsion eaten away and insect carcasses scattered among the records) (Figure 1).

Upon the collection's assessment, it was soon clear that no organisational pattern was in place. This is not uncommon: because photographs are chemically and physically fragile, they are often removed from other text-based media in order to ensure appropriate housing. Photographs thus frequently lack a discernible original order or clear provenance that usually inform best practice procedures in terms of arrangement.[10] This structural disorder, however, was primarily due to institutional abandonment and amnesia. No documentation related to its making, use, or custodial transfer existed and a substantial portion of its contents was discovered lying loose in large cardboard boxes (Figure 2). The items had been moved many times over and thus lacked contextual details key to empirical historical investigation.

The UCT DoS Collection was thus recognised as 'orphaned' in that it suffered from a lack of provenance, its origins remained uncertain, and its place within its custodial institution was deemed precarious.[11] It thus troubled typical archival methods and protocols. Most pressingly, as medical

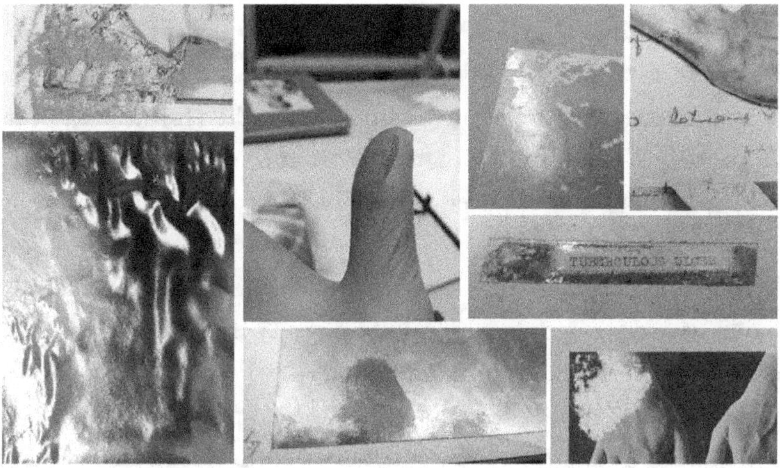

Figure 1. The physical state of photo-objects in the UCT DoS Collection as discovered during its initial assessment. The material shows evidence of long-term neglect, the effects of sub-optimal storage environment (fluctuating temperature and humidity), and a lack of pest control.

Figure 2. A portion of the UCT DoS Collection is shown as it was discovered in a storage room at the Department of Surgery of Cape Town's medical school before being transferred to the UCT Pathology Learning Centre.

material produced in a colonial context and without clear custody, questions about private data, sensitive subject matter, and possible disenfranchisement all accompanied questions of uncertain image rights. It thus had what Diane Vogt-O'Conner calls a "high maintenance mortgage" – it posed too many problems and required too many resources to make it viable for research purposes.[12]

Medical photographs are frequently 'unearthed' or 'rediscovered' in this way, often as abandoned documents hiding in "neglected storage closets, tucked behind file cabinets, or hanging on walls."[13] This is likely due to the general consensus that medical institutions (especially hospitals) often forgo the appropriate preservation of their records due to lack of storage space, instead opting to destroy that which is no longer deemed relevant for short-term use.[14] The result is that the historical context of materials that *are* retained often is lost and nearly impossible to reconstruct.

'Detached' from their origins, material of this kind offers a problem for both historical research and archival practice.[15] Context is vital to maintaining the research value of a collection; and any infringement or loss thereof is also a loss of "evidence and authenticity."[16] As the general goal of historians is to make sense of and find explanations for developments over time, absences pose a significant problem.[17] Orphaned objects or records thus require effort to restore their evidentiary value as primary source material.

Provenance research offers a solution to this problem and is widely adopted by museums containing materials of uncertain origins.[18] In brief, this process begins by studying the characteristics of the object as well as its place within a collection (its relationship to other associated items).[19] It is an interdisciplinary and necessarily versatile method that relies on making connections across items, collections, institutions, and even national borders.[20] The first port of call is the item itself and involves a search for physical properties, visual clues, or legible annotations. Going further "is a step-by-step working process" that sees "different pieces of information linking together" and requires "a combination of art-historical, historical, biographical and genealogical research" to provide hints of an item's journey through time and space.[21] The use of writing media (pencil vs.

ink, stamps, the use of colour, handwriting style, and the addition of typed labels) as well as paper type (quality, colour, texture, and degree of deterioration) may all offer contextual leads.[22] Following up on these leads requires lateral thinking, a willingness to cast a wide net of inquiry, and a readiness to follow traces and hunches that may lead to dead-ends.[23]

Work of this kind makes use of significant resources in terms of time, manpower, and money.[24] However, to research provenance is to perform due diligence.[25] As "custodians, not owners, of cultural information,"[26] heritage institutions have an obligation to be accountable to society at large.[27] This is particularly pressing in relation to colonial collections where provenance is often "fragmentary and ambiguous"[28] as acquisition records were frequently discarded, lost, or destroyed if they ever existed at all.[29] As much colonial material was used to reinforce and circulate racist ideas in the sciences as well as in the public sphere, provenance research is not only an intellectual but an ethical and political act.[30] It serves "as [...] a critical project [...] generating new forms of understanding objects and the social relationships surrounding them."[31] Or, in the words of Ciraj Rassool, it is "an epistemic project, a means of remaking museums and rethinking the relationship between museums and society."[32]

Researchers of orphaned archaeological artefacts have shown that unprovenanced materials lacking contextual information can be re-situated through the practical steps of object management referred to as 'curation'. As Barbara Voss and Megan Kane as well as Zanna Friberg and Isto Huvila demonstrate, physical and intellectual processing (handling, ordering, cataloguing, and caring for material) can constitute a generative research activity in and of itself.[33] Curation's close, item-level, object-centred engagement has the potential to lead to further inquiries, making it a 'practice-based' activity: it is a mode of research that sees the investigator play a constitutive role in establishing connections through practical action and through direct, first-hand experience 'on the ground.'[34] The "sensual experience of handling and caring" is thus key to curation as a research method.[35]

But this is also key to archiving. Making an archive (the processes of selection, ordering, description, and care) relies on the interpretation and decision-making of individuals.[36] The physical management of records during the processing of archival records "builds understanding" and "leads to ready recognition of effective […] strategies."[37] The archivist has an "active presence in making sense of the traces remaining, in shaping selections, and in using analytical ideas to produce theoretical and other interpretations."[38] While archival standards suggest that there is a degree of objectivity in the management of historical material, there is often a heavy reliance on instinctive or 'tacit' knowledge—practical awareness and skills learnt through trial-and-error that becomes "imbedded in the human body and brain."[39] Curation places emphasis on this practice-based dimension of archiving.

Following the tenet of provenance research – that "[t]he collection item itself […] can help illuminate the context of its provenance"[40] – the photographs in the UCT DoS Collection were taken as the first port of call. The archiving of this material (including arrangement and description) was directly informed by the process of the primary physical and visual appraisal. This choice relates to Voss' conviction that artefacts have "the capacity to 'kick back' against applied ideas, models and theories" —allowing for a grounded engagement with and treatment of this colonial medical material.[41]

To establish physical control over the UCT DoS Collection, the decision was taken to arrange it as an 'artificial' collection: a group of items "assembled and organized to facilitate its management or use" and "arranged for the convenience of description or retrieval rather than in an order originally established by the creator."[42] This is not atypical in the case of private collections or repositories of personal papers where it may be necessary "to impose some kind of order on a random collection of material."[43] To do so required a period of record-based analysis to come to grips with its contents. This "micro- or bottom-up approach"[44] meant handling and studying every item individually due to the uncertain and variable nature of the photo-objects (their material make-up, chemical structure, format, and size).[45] In addition, it offered an opportunity to implement basic preservation practices to ensure that the physical integrity of the records could be maintained.[46]

In working through the UCT DoS Collection, each photographic card was scrutinised in terms of its subject matter (was it of a patient, a specimen, or an x-ray image), the visual language (lighting, pose, composition), and any annotations. Every item was lightly dusted to remove surface-level contaminants, and any metal clips, rubber bands, or other volatile materials were removed. Item-level reference numbers were assigned to provide initial physical and intellectual control over the collection.[47] General observations and stand-out content were photographed and noted. This was used to produce blog posts that served to summarise, trace, and hold a public record of the initial archival processing. Such recording or 'mapping' is not only in line with archival best practice, it is also recommended when conducting provenance research in order to follow the research steps as well as ensure accountability.[48]

Initial investigations indicated that the photographs in the collection did not remain solely within the confines of the medical school. Markings on the material (such as arrows, measurements, and scribbled areas) hinted at publication marks for cropping, resizing, and numbering (Figure 3). Marks of this kind operate "like stamps in a passport"[49] – offering glimpses of the photograph's social life. For instance, large letters marking "N'YANGA" and "INYANGA" (an Ndebele word for 'healer') and a stamp reading "Clinical Proceedings" respectively led to the journals *Cathartic* and *Clinical Proceedings*.[50] Photographs from the UCT DoS Collection were identified through visual recall, most notably by recognising specific patients (sometimes heavily cropped) as well as subtle stylistic or formal attributes (such as a signature downward tilt of the camera or lighting techniques) in these publications. Recognition was intuitive and based largely on hard-won familiarity with the visual contents of the collection. The tacit knowledge gained while working through the collection provided a 'trained', 'expert', or 'calibrated' eye for identifying published versions of the material.[51]

Digital and physical searches in local publications, including The *South African Medical Journal*, the *South African Journal of Surgery*, the *South African Journal of Clinical Science*, and the *Central African Journal of Medicine*, revealed that much of the imagery featured in the UCT DoS Collection travelled widely. Some photographs had been reproduced many times over, in multiple articles by different authors and across publications as well as time periods (Figure 4). The presence of material from the UCT DoS Collection

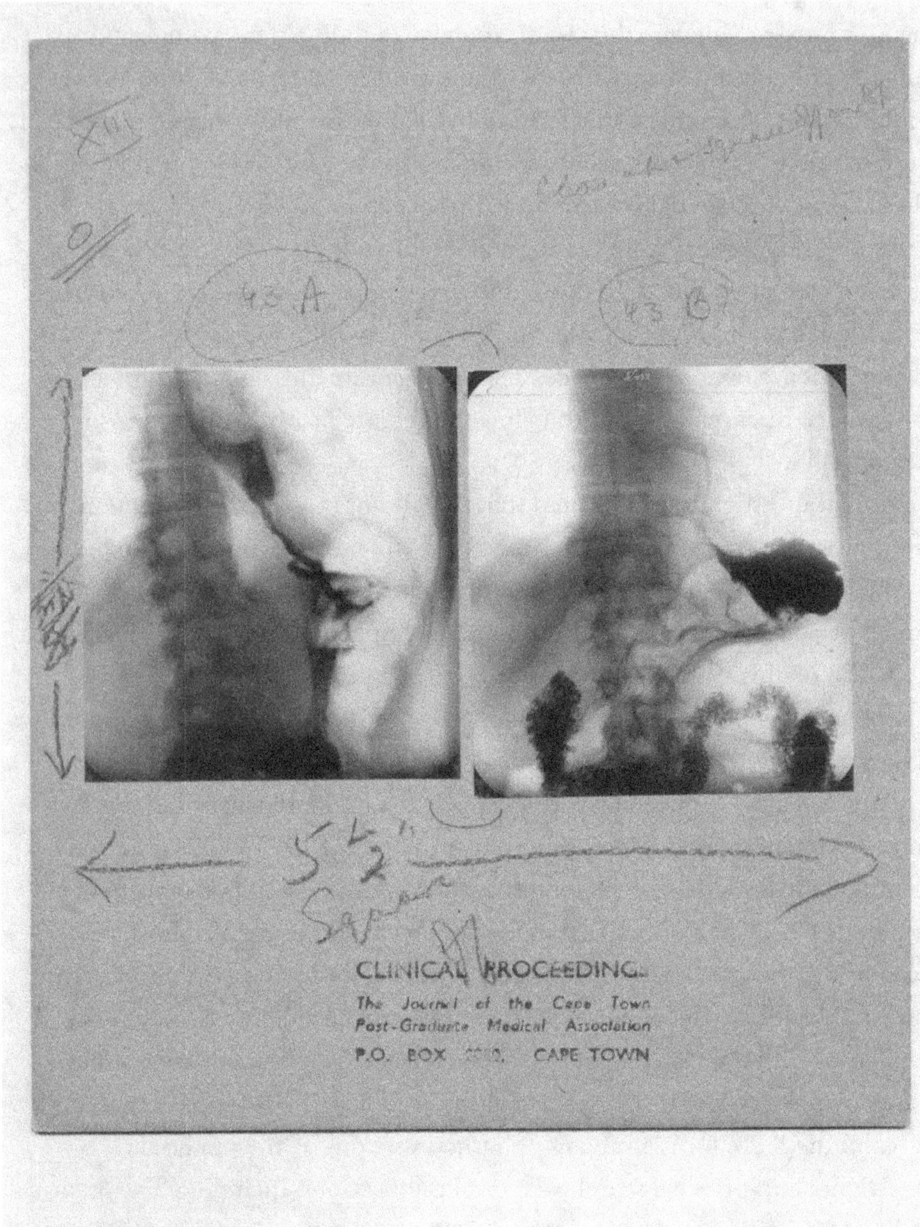

Figure 3. A photo-object intended for publication in the journal 'CLINICAL PROCEEDINGS' as indicated by a stamp. Each x-ray photograph is identified according to how it would be labelled in the journal article ('43A' and '43B'). The addition of hand-drawn arrows indicates how much of the images is to be reproduced and how, together, they should be resized for this purpose ('5½" Square').

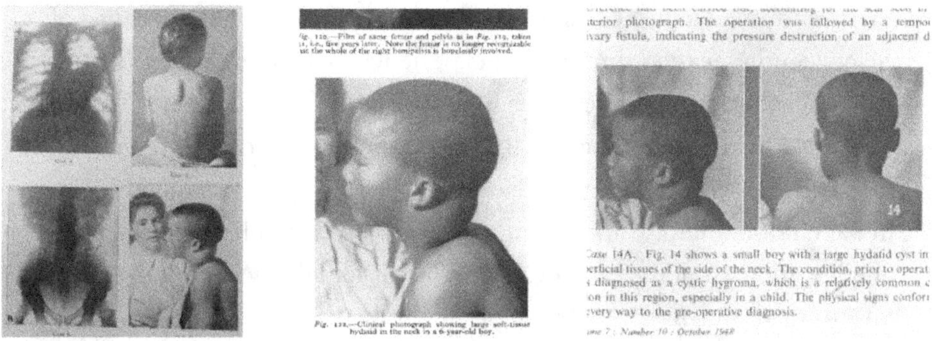

Figure 4. Examples of the wide reuse of photographs from the UCT DoS Collection can be seen here. The same photographs of a patient with a significant swelling of the neck are reproduced in M Horwitz's 1943 article on Hydatid Disease in *Inyanga* (left). These were to be used again by CFM Saint in a 1948 article on Hydatids published in *Clinical Proceedings* (centre) as well as by W James Latham in a 1953 article on the same topic published in *The Faculty Journal of The of Radiologists* (right).

within these publications confirmed that the photo-objects were not merely used for teaching – their visual content also served as illustrative material for more public discussions of pathological cases and clinical care.

To log attributes of the photo-objects, visual and textual details were entered into an electronic database to serve as an accession record and offer a means to make each item discoverable.[52] This also served to translate the collection's visual and material attributes into writing. Initially, the points of entry were fairly standard with regards to photographic description.[53] They included basic fields including index number, date of processing, the genre of the photograph (specimen, patient, x-ray, or other), the colour of the mounting board, the number of photographs on the individual mount, the most prevalent bodily region imaged, and any information written on the photo-object itself (frequently a date, details related to the patient and their diagnosis). A field was added to calculate whether access to the photo-object should be restricted based on the Wellcome Collection's recommendation to assume a 100-year lifespan for subjects depicted in clinical photographs.[54]

As this process highlighted, the act of classifying and cataloguing records involved active decision-making informed by tacit knowledge of the collection gleaned through the experience of working with the records themselves.[55] Handling the collection's contents revealed the need for the additional fields while some existing categories required a degree of inference in order to enter relevant data. For instance, while a patient's name and age are frequently annotated on the verso of the photographic cards, whether they would be considered 'male' or 'female' never appears. Reference to race was similarly elusive despite the records having been produced in a highly racialized socio-political environment.[56] Annotations designating racial categories were rare,[57] the mounting boards did not suggest race-based colour-coding, and the monochromatic nature of the images made race hard to 'read' in the images due to the shifting tone and inconsistent contrast within and across the photographs.[58] 'Race' was thus omitted as a data field; instead the 16 cases in which race-based terminology was used (with terminology including 'native' and 'coloured', 'Cape Coloured', and 'CM' for 'Coloured Male') were entered into the 'notes' section to ensure their discoverability.

The database offered a means to come to terms with the material as a whole by creating what Gesa Grimme calls a 'collection profile' – a broad data-driven overview of a collection's contents and context.[59] The items within the UCT DoS Collection were tabulated in relation to year and translated into graphic form in order to gain a scope of the over 60-year period (Figure 5). It was hoped that this chronological ordering would reveal changes over time in terms of the collection's visual and material make up and thereby offer glimpses into technological, institutional, and social changes related to clinical photography and medicine in Cape Town. There remained, however, many aspects that could not be so easily tabulated including subtle, unquantifiable visual and material features. While the database was a great assist in later research, it was working through the photo-objects that served as the most important attribute in the collection's curation.

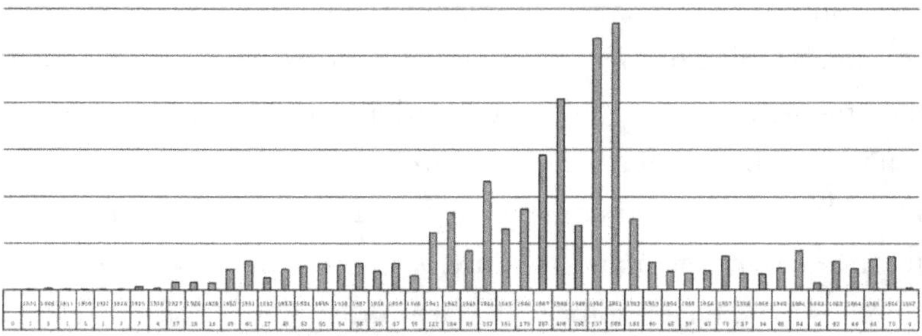

1900s		1910s		1920s		1930s		1940s		1950s		1960s	
						1930	45	1940	28	1950	535	1960	48
1901	1	1911	1			1931	60	1941	119	1951	569	1961	84
				1922	1	1932	25	1942	158	1952	152	1962	16
						1933	44	1943	82	1953	61	1963	62
				1924	2	1934	52	1944	223	1954	40	1964	46
				1925	7	1935	55	1945	129	1955	28	1965	68
1906	3			1926	4	1936	54	1946	171	1956	37	1966	70
				1927	15	1937	58	1947	282	1957	87	1967	4
				1928	18	1938	40	1948	408	1958	35		
		1919	1	1929	15	1939	57	1949	139	1959	36		
	4		**2**		**62**		**490**		**1739**		**1580**		**398**

Figure 5. Graphic and tabular overview of UCT DoS Collection drawn from the Microsoft Access database. The graph shows the number of photo-objects dated by year and in chronological order. This offers a rough indication of the quantity of photographs produced on an annual basis. The table was manually created and shows the number of photo-objects dated by year and grouped by decade.

Material Rewards

The UCT DoS Collection is made up of stand-alone items imbedded in individual textual contexts (their mounting cards with annotations). Notations on the material confirmed that the majority of the collection's contents were made at Cape Town institutions by mention of local facilities, medical practitioners, and publications. Mounts prevent direct contact between the body of the researcher and the photograph,[60] thus allowing for 'legitimate' tactile engagement with the photograph as a fragile object.[61] The mounting is thus indicative that the material was made to be rifled through, turned over, and passed between persons while keeping the image safe.[62] Housed in a clinical department at Cape Town's medical school, it appeared likely that the cards in the collection were mounted in order to serve as visual teaching aids that could be touched, shared, and inspected by students of surgery.

The contextually imbedded nature of the UCT DoS Collection provided a concrete means to relate the records to one another chronologically and thereby highlight changing visual and material attributes over time. While invisible to the database, tactile and aesthetic features aided in periodising items when other contextual details were missing. For instance, the shifting colour, weight, and texture of the mounting board offered an indication as to when a photograph was produced, while the date-based organisation allowed for individual case histories to emerge.

Grouped according to date, clinical photographs (images of the patient), x-ray photographs (reproductions of x-ray images), and specimen photographs (depictions of excised organs) related to a single patient were brought together to reconstruct a patient's medical experience. When dates did not perfectly match or annotations (such as the patient's name) were missing, it was the tacit knowledge and familiarity with the collection that allowed for the grouping together of case histories whose connections might otherwise have been missed. For example, in some instances the same patient may have been imaged from different angles or on a different day (such as x-rays or post-operative specimen photographs). Here it was only by recognising the patient's appearance, unique diagnosis or severe symptom that these photo-objects could be connected.

Undated items also offered insight into the varied sources of items within the UCT DoS Collection. One stand-out, photo-object depicts a patient showcasing the surgical scarring left having undergone a hip replacement and annotated as "Arthroplasty of hip." Embossed upon the bottom right corner is the title of a commercial photographic studio, Vernon Studios located in Cape Town's Plein Street—a bustling urban hub that hosted many businesses and boasted a thriving commercial nightlife from the mid-nineteenth century on.[63] Globally, the use of the space and skill of a studio photographer was not uncommon during the late nineteenth and early twentieth centuries as the photographic medium required a cumbersome suite of apparatuses and knowledge of specialist technical chemical processes. World-wide, professional photographers or commercial studios were frequently called upon to render their services to medical practitioners.[64] Many early clinical photographs thus demonstrate an aesthetic overlap with the commercial photographic studio. In this example, the patient's stance, the framing, the lighting, and the hints of a draped backdrop are visually reminiscent of nineteenth century *Académies* – nude photographs mimicking established artistic tropes including the active stance of an athlete or hero.[65] In this clinical photograph, the theatrical pose of the patient suggests the influence of wider cultural traditions.

Arranging the UCT DoS Collection chronologically revealed that the photographs evolved in terms of their visual language over time. Material up until the latter half of the 1930s is notably varying in depictive practices, with patients frequently photographed against various backdrops (both indoors and outside), wearing their own clothes, captured from divergent angles, and variously positioned within the photographic frame. Contrast, lighting, focus, clarity, grain, paper, and developing techniques are inconsistent and frequently see items suffer from different forms of age-related deterioration (such as silvering, fading, or cracking). Photo-objects dated to the 1920s and the first half of the 1930s demonstrate patients unbuttoning, hiking up, or otherwise maneuvring their clothes to reveal the afflicted area of their body. Together, such features suggest that photographs were taken in an institutional environment that lacked both a formal space for photographic work as well as a designated private location for patients to undress.

Visual and material evidence further hints at technological shifts. Some of the loose (unmounted) photographs are suggestive of the photographic technology used during this early period. Two photographs see the printing paper cut wider than the individual image frame, showing the edges of a second image. This characteristic is indicative of the seriality of roll film—a single strip of film wound behind the camera lens upon each exposure. Unlike a professional camera (that would use separate film sheets), these photographs suggest the use of commercial hand-held equipment such as a 'hand', 'box', 'pocket', or other 'miniature' camera that used film of this kind and was often preferred by non-professional photographers.[66] The physical size and chemical treatment of these examples in the UCT DoS Collection further demonstrate amateurish practices. They have been printed on a 1:1 scale with the film suggesting the process of contact printing—a method whereby the photographic printing paper is placed directly beneath the film in order to create a direct, inverted copy of the negative image. This technique requires no additional or expensive darkroom equipment (such as a projector or 'enlarger') and thus serves as a cost-effective mode of image-making as well as one that requires minimal specialist knowledge. The prints also suggest a large degree of inexperience both with a camera and in the darkroom as the presence of dark fingerprint stains indicates messy chemical use and careless processing practices.

Other examples suggest experimentation in photographic processing and printing. There is evidence of (rather unsuccessful) attempts to create stereoscopic images to create the illusion of three-dimensional vision (when seen through a viewfinder or 'stereoscope'),[67] as well as the addition of a vignette (the oval frame typical of Victorian portraiture) that would often be a standard darkroom accessory but does not appear to serve a medical purpose. Material choices too suggest a large degree of both improvisation and experimentation. The use of highly textured cream coloured photographic paper seems wholly inappropriate for clinical purposes as the rough grain can interfere with visual detail by subtly distorting the emulsion layer and thus the surface of the image.[68] There are also instances where this paper appears to have been used in the printing of negative versions of some of these photographs, creating a distinctive textured pattern in which the thick weave prevented an even print.

Together, the variations in material make-up, locational and visual attributes, and aesthetic as well as functional choices discussed above demonstrate highly inconsistent and even amateurish photographic practices. The technically 'unprofessional' appearance of many of the photo-objects dated to the first decades of clinical education at Cape Town's medical school suggests that clinical photography was not practiced by professional photographers or commercial photographic services but rather made in an informal and *ad hoc* manner.

To identify possible historical actors involved in these practices was, once again, facilitated through the photo-objects themselves. Key items in the UCT DoS Collection showed evidence of having originated beyond the borders of South Africa. One depicts an excised portion of the digestive tract, "Ileum Caecum & Ascending Colon," and includes a lengthy handwritten inscription referencing the removal of the organ from a 36-year-old a patient in 1912 at Newcastle-on-Tyne's Royal Victoria Infirmary. Crucially, it identifies by name a James Rutherford Morison (1853-1939) – a pioneer in British surgery who worked as both surgeon (at the Royal Infirmary) and surgical lecturer (at Durham University College of Medicine). While there is no record of Morison ever having visited Cape Town or its medical school, he was mentor to a key figure associated with this facility.

Charles Frederick Morris (CFM) Saint had studied and worked under Morison in Newcastle-on-Tyne before becoming the first head of Surgery in Cape Town in March of 1920 at the age of 33.[69] Widely held to be South Africa's "father of surgery,"[70] Saint appears to have physically brought some of the items within the UCT DoS Collection with him from England as exemplified by not only Morison's photograph but by two cabinet cards[71] produced by a self-proclaimed 'Art Photographer,' RE Ruddock, in Newcastle-on-Tyne's Goldsmiths Hall. (Figure 6) Further, multiple photographs featured in the UCT DoS Collection are either original prints or halftone reproductions of Morisons's widely publicised surgical textbook *An Introduction to Surgery* (1910). (Figure 7)

Once in Cape Town, Saint made moves to introduce clinical photography to his newly founded surgical department on South African soil. Having converted a small cupboard into a dark-room on the original campus of Cape Town's medical school,[72] Saint appears to have done what many medical facilities did at the time – relied on the 'semi-skilled' labour of a technician to fulfil photographic work.[73] The person in question was Ernest James Henderson, the Department of Surgery's "first-grade technical assistant and general handyman."[74] While this figure is invisible within the UCT DoS Collection, 37 photo-objects within it were published in later editions of Morison's textbook (co-authored by Saint) with an acknowledgement of Henderson for some of the "excellent photographs" it featured.[75]

Hired as a technician in the 1920s, Henderson would probably not have had any formal training in photography. Internationally, the early twentieth century saw technicians often employed at a young age, without any work experience, and sometimes directly out of school. Charged with menial tasks

Figure 6. A cabinet card depicting a young patient with a significant protrusion from the forehead. The photograph was produced by the commercial photographer RE Ruddock of Newcastle-on-Tyne and is evidence of the use of a commercial photographic studio to produce clinical images.

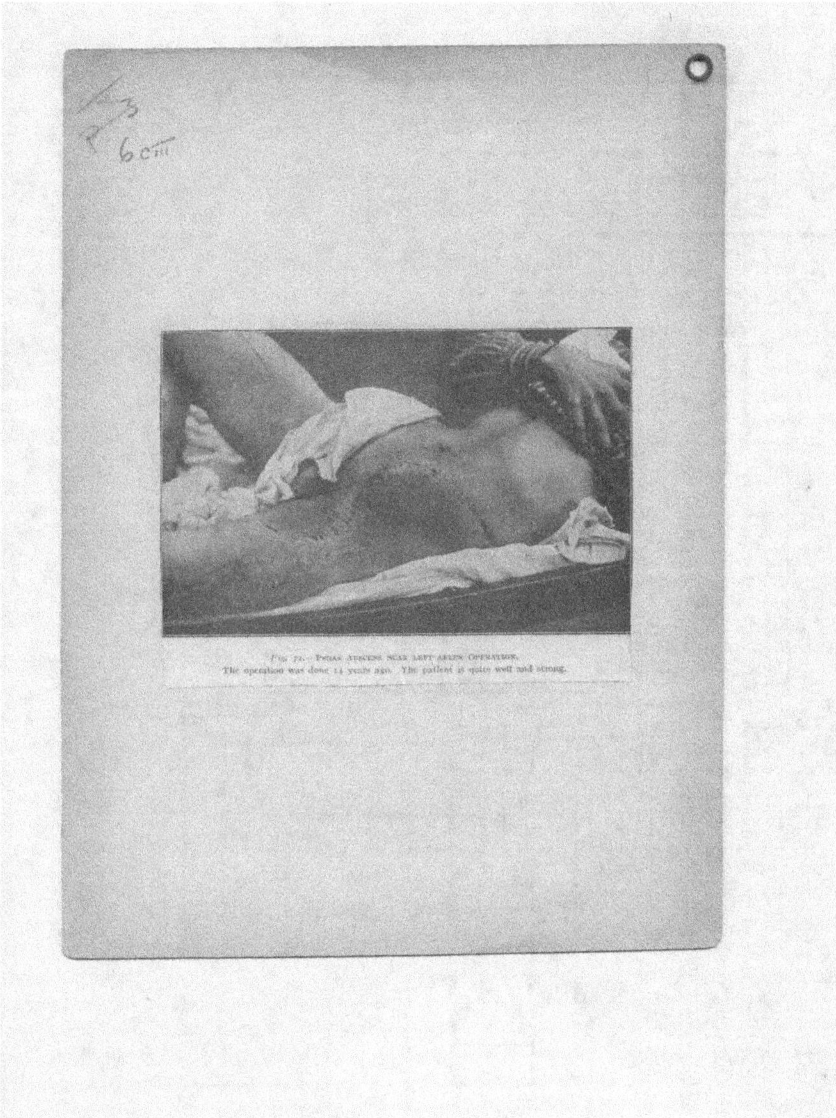

Figure 7. Published halftone print of a photograph depicting a patient with a significant post-operative abdominal scar. This reproduction originated from the second edition of Morison's textbook *An Introduction to Surgery* (1925), co-authored with Charles Frederick Morris Saint. The original print is in the UCT DoS Collection, suggesting that the photo-object was produced overseas (in Newcastle-on-Tyne) and physically brought to Cape Town, rather than printed and mounted locally in South Africa.

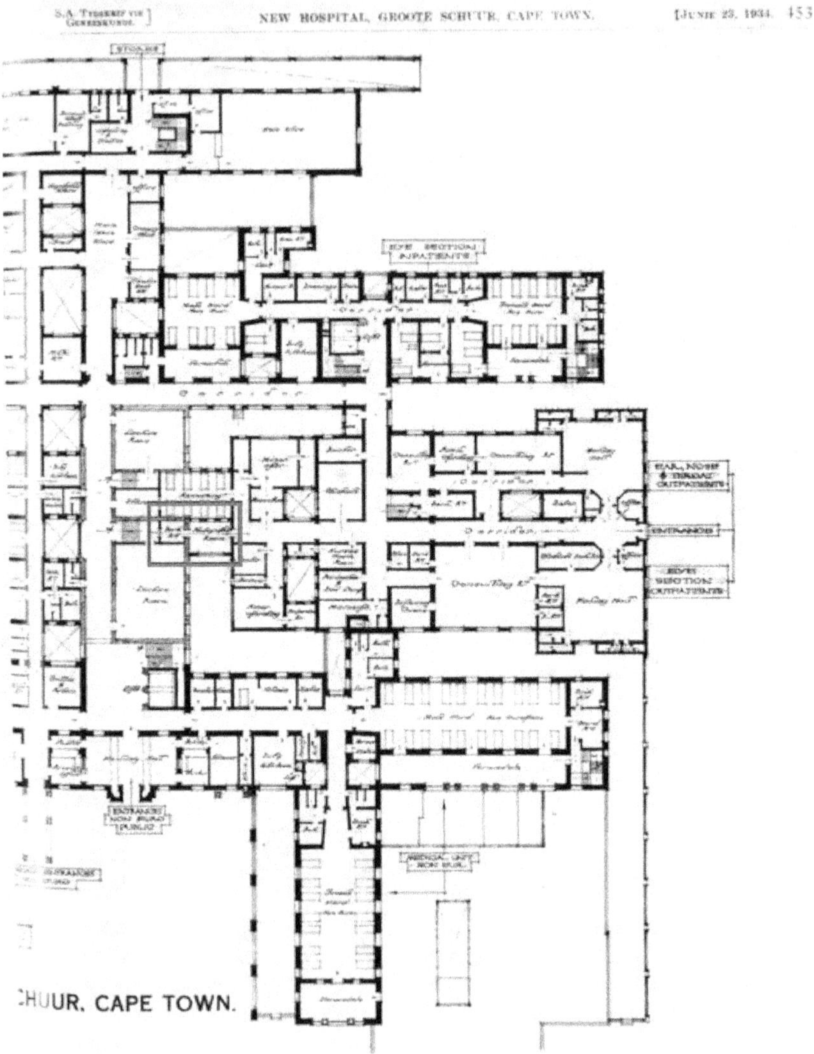

Figure 8. Detail from 1932 floorplan of GSH published in the *South African Medical Journal.*

and trained piecemeal, a technician's skillset, knowledge, and role was thus heavily reliant on the needs of the division for which they worked or the individual staff member who would groom them into a "bespoke right-hand man."[76] Henderson (like other technicians at the time) would have reported directly to his division head and would have been recruited to incorporate photographic work into his repertoire.[77] The informal, experimental, and inconsistent photographic practices evident in the UCT DoS Collection at the time of Henderson's tenure confirm these suspicions.

Towards the end of the 1930s, it appears that clinical photography at Cape Town's medical school underwent a process of standardisation. Increasingly, the aesthetics of the photographs in the UCT DoS Collection as well as visible traces of the photographic space hint at the use of a formal studio and equipment as well as the application of professional photographic procedures. Photographs that centre the patient's head, neck, shoulders and torso frequently depict two or more points of view (full-frontal, profile, and sometimes angled). The decontextualizing and neutralizing procedures of angle, focus, background, and dress serve to focus on the symptomatic site. Attention is clearly placed on representing the bodies of patients in a consistent manner, rendering the images (and the patients as well as their conspicuous symptoms) comparable.

Photographs depicting lower limbs of patients indicate the use of a (otherwise unseen) piece of photographic equipment – namely a tripod – that can be inferred by the angle at which the lower portion of the body is captured. While photographs of heads and busts are rendered consistently from an eye-level viewpoint, the perspective granted in photographs of legs and feet denotes of the camera having been tilted downward. Another technology indirectly evidenced is artificial standing lights made visible by the even, shadowless lighting of patients (on the one hand) and the use of deliberate shadows cast by a single controlled light source (on the other). While the former seeks to flatten the image and remove the distraction of irrelevant depth and non-clinical detail, the latter is harnessed to accentuate the extent of bodily distortion.

Photographs dated from the late 1930s further show that the patients imaged were confined to a single, designated room established and equipped for clinical photographic work. Signature architectural features including unique wooden panelling, metal hinges, and a tell-tale handle can be glimpsed on the peripheries of photographs depicting upper bodies, torsos, and heads. The spatial fixity visible in these examples from the UCT DoS Collection coincides with the physical move of clinical teaching to a brand new facility, Groote Schuur Hospital (GSH), in 1938.[78] A hospital plan (published in 1932) reveals an institution split down the middle into 'European' and 'Non-European' wards – a racial divide that would be implemented and policed until the 1980s.[79] Accommodated too was a "Photographic Room" and a "Dark Room" that would, at least according to the evidence in the photographic collection, cater to individuals across demographic groupings (Figure 8). In short, while the hospital as a whole was segregated along lines of race, men and women, young and old, and individuals with various skin tones (and thus spanning racial designations) were imaged within the same studio.

The opening of a professional photographic space at GSH and thus the establishment of infrastructure for clinical photography (the room, backdrop, tripod, lighting) offers an explanation for some of the aesthetic developments in the UCT DoS Collection. However, the standardized imaging protocol (the photographic methods applied) and thus the application of professional photographic technique was due to the introduction of a new staff member to the Department of Surgery. With Henderson's retirement in 1940 came Gerald Richards McManus. Having begun his career as a 'learner technician' in 1918 at the age of 16 in the Department of Bacteriology, McManus shifted to Surgery as a 'technical assistant' in 1940.[80] McManus was thanked by Saint "for [his] help with the photography"[81] in the fourth edition of Morison's *An Introduction to Surgery* (1948) and is repeatedly acknowledged for his photographic contributions in numerous articles published in the *South African Medical Journal*. A year after McManus' transfer to the Department of Surgery, the UCT DoS Collection shows a significant spike in contents – jumping from 30 photo-objects in 1940 to 122 the following year. McManus thus appears to have become Saint's new 'right hand man' with regards to clinical photography, adding material to the UCT DoS Collection until 1967. It is in this year that McManus retires; it is also at this time that the last photo-object is added to the collection.

Conclusion

This chapter has sought to outline the method of curation as a means to reinstate the research potential of the UCT DoS Collection. In the successful application of visual and material methods, the chapter traced the history of clinical photography at Cape Town's medical school from the first imported records to the gradual professionalization of local photographic work. Following the photographs further allowed the interlocking networks of British surgeons, local technical assistants, and international publishing practices to be untangled. As such, the analysis testifies to the productive possibilities of decontextualized clinical collections for the historian of medicine.

Processing the collection through experiential and grounded engagement had the distinct advantage of forcing close, repeat encounters with the photo-objects. This, ultimately, equipped researchers of the material to "knit together matter and meaning" – to allow for threads of intellectual inquiry to be driven by both visual and physical evidence.[82] In taking the photographs as the first port of call, the handling and researching of this material were performed as concomitant endeavors, thereby taking up Voss' advocacy for curation as a solution to the problem of orphan works. Here, the deviation from traditional archival methodology was in response to the unique circumstances of the collection as a repository of decontextualized photographic material produced within the context of colonial medicine.

Endnotes

1 Barbara L. Voss, "Curation as Research: A Case Study in Orphaned and Underreported Archaeological Collections," Archaeological Dialogues 19, no. 20 (2012): 145-169; Barbara L. Voss and Megan S. Kane, "Re-establishing Context for Orphaned Collections: A Case Study from the Market Street Chinatown, San Jose, California," *Collections: A Journal for Museum and Archives Professionals* 8, no. 2, (2012): 87-112.

2 Elizabeth Edwards, "Material Beings: Objecthood and Ethnographic Photographs," *Visual Studies* 17, no.1 (2002); Elizabeth Edwards and Janice Hart, *Photographs Objects Histories: On the Materiality of Images* (London & New York: Routledge, 2004); Elizabeth Edwards and Siegried Lien, *Uncertain Images: Museums and the Work of Photographs*, (London & New York: Routledge, 2014).

3 Igor Kopytoff, "The Cultural Biography of Things: Commoditization as Process," in *The Social Life of Things: Commodities in Cultural Perspective*, ed. Arjun Appadurai (Cambridge: Cambridge University Press, 1986), 64-91.

4 Arjun Appadurai, ed., *The Social Life of Things: Commodities in Cultural Perspective* (Cambridge: Cambridge University Press, 1986).

5 Costanza Caraffa, "Objects of Value: Challenging Conventional Hierarchies in the Photo Archive," in *Photo-Objects: On the Materiality of Photographs and Photo Archives in the Humanities and Sciences*, ed. J. Bärnighausen, Caraffa, C, Klamm, S, Schneider, F, and Wodtke, P. (Berlin: Max Planck Institute for the History of Science, Studies 12, 2020), 14-15.

6 Elizabeth Edwards, *Raw Histories: Photographs, Anthropology and Museums* (Oxford: Berg, 2001), 2.

7 Barbara Lazenby Craig, "The Canadian Hospital in History and Archives," *Archivaria* 21 (1985-86): 52-67; Elizabeth Denham, "Dealing With the Records of Closing Hospitals: The Calgary Area Health Authority Plan," *Archivaria* 41 (1996): 78-87; Hamish Maxwell-Stewart and Alistair Tough, "Cutting the Gordian Knot: Or How to Preserve Non-Current Clinical Records Without Being Buried in Paper," *Archivaria* (1996): 61-77; Voss, "Curation as Research;" Paul Valentich-Scott, *Saving Orphan Collections*, Webinar for Small Collections Network (2015); Tiffany Adrain and Paula Work, *SPNHC: Threatened and Orphaned*

Collections.

(2017); Elizabeth Edwards, "Location, Location: a Polemic on Photographs and Institutional Practices," *Science Museum Group Journal* (2017).

8 Kate Arnold-Forster, "'A developing sense of crisis': a new look at university collections in the United Kingdom," *Museum International* 52, no. 3 (2000): 10-14; Peter Stanbury, "University museums and collections," *Museum International* 52, no. 2 (2000): 4-9; Nick Merriman, "University museums: problems, policy and progress," *Archaeology International* 5 (2001): 57-59; Nick Merriman, "The current state of Higher Education Museums, Galleries and Collections in the UK," *Museologia* 2 (2002):71-80; Andrew Simpson, "Cinderella Collections, fifteen years after the ball: Australia's university museums reviewed," *Museums Australia Magazine* 21, no. 2 (2012): 18-20; Panu Nykänen, "First steps in global advocacy: some perspectives on the formation of UMAC, an international committee of ICOM," *University Museums and Collections Journal* 10 (2018): 10-21.

9 The UCT DoS Collection falls into the category of 'University Museums and Collections' (UMAC) that frequently see changes in institutional priorities, shifting methods of teaching and learning, financial constraints, the lack of clear collections policies or specialised staff, and the destruction, merging, or development of university facilities coupled with the growing emphasis on digital (rather than material) resources all contribute to institutional vulnerability. Arnold-Forster, "A developing sense of crisis;" Stanbury, "University museums and collections;" Merriman, "University museums;" Merriman, "The current state of Higher Education Museums;" Simpson, "Cinderella Collections;" Nykänen, "First steps in global advocacy."

10 Carnell, "Accessioning and Arrangement," 145-146.

11 Steven W.G. de Clerq, "Uniting Forces: The European Network and National Collaborative Projects," in *Managing University Museums: Education and Skills*, (Paris: OECD, 2001), 85-102; Caroline Williams, *Managing Archives: Foundations, Principles and Practice* (Oxford: Chandos Publishing, 2006); Richard M. Leventhal and Brian I. Daniels, "'Orphaned Objects', Ethical Standards, and the Acquisition of Antiquities," *DePaul Journal of Art, Technology & Intellectual Property Law* 23, no. 2 (2013): 339-361; Valentich-Scott, *Saving Orphan Collections*; Adrain and Work, *SPNHC*; Nykänen, "First steps in global advocacy."

12 Diane Vogt-O'Connor, "Appraisal and Acquisitions," in *Photographs: Archival Care and Management*, ed. Mary Lynn Ritzenthaler and Diane Vogt-O'Connor (Chicago: Society of American Archivists, 2006), 155.

13 Jeffrey Mifflin, "Visual Archives in Perspective: Enlarging on Historical Medical Photographs," *The American Archivist* 70, no. 1 (2007): 62.

14 Craig, "The Canadian Hospital," 56; Rochelle Keene and Julie Parle, "Museums, Archives and Medical Material in South Africa: Some Ethical Considerations," *South African Museums Association Bulletin* 37 (2015): 6-16.

15 Williams, *Managing Archives*, 74.

16 Ibid., 74-75.

17 Anne Green and Kathleen Troup, "*The Houses of History: A Critical Reader in History and Theory*, 2nd Edition, (Manchester: Manchester University Press, 2016), 16; Leonie Hannan and Sarah Longair, *History Through Material Culture*, (Manchester: Manchester University Press, 2017), 58-59.

18 Provenance research originated in the domain of art history and was primarily employed to prove the authenticity of an artwork. However, the 21st century has seen the use of the term expand to encompass collections and historical items more broadly. Arthur Tompkins, *Provenance Research Today: Principles, Practice, Problems* (London: Lund Humphries, 2020).

19 Jonathan Fine and Hilke Thode-Arora, "Provenance Research – Research Sources, Methods, and Possibilities," in *Guidelines for German Museums: Care of Collections from Colonial Contexts*, 3rd edition (Berlin: German Museums Association, 2021), 147.

20 Sharon Flescher, "The Challenges of Provenance Research," in *Provenance Research Today: Principles, Practice, Problems*, ed. A. Tompkins (London: Lund Humphries, 2020); Marie Stolberg and Andrea Lehmann, "Best-Practice Guidelines, Research Methods and Tools," in *Provenance Research Today: Principles, Practice, Problems*, ed. A. Tompkins (London: Lund Humphries, 2020).

21 Stolberg and Lehmann, "Best-Practice Guidelines," 55.

22 Ibid., 55-56.

23 Ibid., 67.

24 Fine and Thode-Arora, "Provenance Research," 147.

25 Fleschner, "The Challenges of Provenance Research;" Freda Matassa, *Museum Collections Management: A Handbook*, (London: Facet Publishing, 2011); Tristram

Besterman, "Museum Ethics," in *A Companion to Museum Studies*, ed. Sharon Macdonald (Oxford and Victoria: Blackwell, 2006), 431-441.

26 Matassa, *Museum Collections Management*, 37.

27 ICOM, *ICOM Code of Ethics for Museums* (Paris: International Council of Museums, 2017), 7-9; American Alliance of Museums, "Ethics, Standards and Best Practices for Museums," in *Alliance Reference Guide* (Washington: American Alliance of Museums: 2013), 3.

28 Larissa Förster, "Plea for a more systematic, comparative, international and long-term approach to restitution, provenance research and the historiography of collections," *Museumskunde* 81 (2016): 52.

29 Fine and Thode-Arora, "Provenance Research," 150.

30 Tony Bennett, "The Exhibition Complex," *new formations* 4 (1988): 73-102; Shelly Errington, *The Death of Authentic Primitive Art and Other Tales of Progress* (Berkeley, Los Angeles, and London: University of California Press: Berg, 1998); Sharon Macdonald, "Exhibitions of Power and Power of Exhibition: An Introduction to the Politics of Display," in *The Politics of Display: Museums, Scien-ce, Culture*, ed. Sharon Macdonal (London and New York: Routledge, 1998), 1-24.

31 Larissa Förster, "Provenance: An essay based on a panel with Ciraj Rassool, Paul Basu, and Britta Lange," in *OTHERWISE: Rethinking Museums and Heritage*, eds Sharon Macdonald and Jonas Tinius (Berlin: CARMAH, 2018), 19.

32 Ciraj Rassool, "Re-storing the Skeletons of Empire: Return, Reburial and Rehumanisation in Southern Africa," *Journal of Southern African Studies* 41, no. 3 (2015): 670.

33 Voss and Kane, "Re-establishing Context;" Zanna Friberg and Isto Huvila, "Using object biographies to understand the curation crisis: lessons learned from the museum life of an archaeological collection," *Museum Management and Curatorship* 34, no.4 (2019): 362-382.

34 Linda Candy, "Practice Based Research: A Guide," *CCS Report* 1 (Sydney: University of Technology, 2006); Steven Pace, "Writing the Self into Resear-ch: Using Grounded Theory Analytic Strategies in Autoethnography," *Text* (2012): 1-15; Carla Willig, *Introducing Qualitative Research in Psychology*, 3rd Edition (Berkshire: Open University Press, 2013).

35 Voss, "Curation as Research," 158.

36 Joan M. Schwartz and Terry Cook, "Archives, Records, and Power: The Making of Modern Memory," *Archival Science* 2 (2002): 1-19; Williams, *Managing Archives*, 52.

37 Brett Carnell, "Accessioning and Arrangement," in *Photographs: Archival Care and Management*, ed. Mary Lynn Ritzenthaler and Diane Vogt-O'Connor (Chicago: Society of American Archivists, 2006), 162.

38 Niamh Moore, Andrea Salter, Liz Stanley, and Maria Tamboukou, eds, *The Archive Project: Archival research in the social sciences* (London & New York: Routledge, 2017), 69.

39 Stuart Blume, *Insight and Industry: On the Dynamics of Technological Change in Medicine*, (Cambridge: The MIT Press, 1992), 231; Henry Collins, *Tacit & Explicit Knowledge* (Chicago: The University of Chicago Press, 2010).

40 Fine and Thode-Arora, "Provenance Research," 147.

41 Voss, "Curation as Research," 149.

42 Carnell, "Accessioning and Arrangement," 148.

43 Williams, *Managing Archives*, 103.

44 Williams, *Managing Archives*, 52.

45 Mary Lynn Ritzenthaler, "Preservation," in *Photographs: Archival Care and Management*, ed. Mary Lynn Ritzenthaler and Diane Vogt-O'Connor (Chicago: Society of American Archivists, 2006), 207-256.

46 Mary Lynn Ritzenthaler, "Integrating Preservation and Archival Procedures," in *Photographs: Archival Care and Management*, ed. Mary Lynn Ritzenthaler and Diane Vogt-O'Connor (Chicago: Society of American Archivists, 2006), 257-270.

47 Ibid., 262; Carnell, "Accessioning and Arrangement," 145.

48 Carnell, "Accessioning and Arrangement," 158; Liz Stanley, "Archival methodology inside the black box: noise in the archive!," in *The Archive Project: Archival research in the social sciences*, Niahm Moore, Andrea Salter, Liz Stanley and Maria Tamboukou (London & New York: Routledge, 2016), 53; Stolberg and Lehmann, "Best-Practice Guidelines," 67.

49 Stolberg and Lehmann, "Best-Practice Guidelines," 55.

50 *Cathartic* was renamed *I'Nyanga* shortly after being launched in 1931 by the Medical Students' Council and then again called *Cathartic* from 1954 until 1973. *Clinical Proceedings* was produced by the Post-graduate Medical Association

(1942-1949) and renamed *Pulse* in 1974. The journal ran under this name until 1990. Jan Hendrik Louw, *In the Shadow of Table Mountain: A History of the University of Cape Town Medical School and its Associated Teaching Hospitals up to 1950, with Glimpses into the Future* (Cape Town: Struik, 1969); Howard Phillips, *The University of Cape Town, 1918-1948* (Cape Town: University of Cape Town Press, 1993).

51 Lorraine Daston and Peter Galison, *Objectivity* (New York: Zone Books, 2010), 44.

52 For the purposes of accessioning the UCT DoS Collection, I chose to use Microsoft Access as it is regularly referenced in archiving guidelines and case studies of small or under-funded repositories. This software has been used in cases where expensive specialist archival software is unavailable or financially untenable. Amanda Stephens, *Archives Management Software Review Report* (Halifax: Council of Nova Scotia Archives, 2008), 63. For more on accession records and discoverability, see Carnell, "Accessioning and Arrangement," 137, 142; Helena Zinkham, "Description and Cataloging," in *Photographs: Archival Care and Management*, ed. Mary Lynn Ritzenthaler and Diane Vogt-O'Connor (Chicago: Society of American Archivists, 2006), 162.

53 Zinkham, "Description and Cataloging," 168.

54 Helen Wakely and Carly Dakin, "Assessing the Sensitivity of Images in Research Collections: A new Approach at the Wellcome Library," *Journal of Visual Communication in Medicine* 38, no. 1-2 (2015): 51-60. A formula in the database uses the patient's date of birth (calculated from the noted age of the patient and the date on the card) in relation to the present date (kept automatically current by the software) to calculate this.

55 Moore et al., *The Archive Project*, 164.

56 Laurel Baldwin-Ragaven, Janelle de Gruchy and Leslie London, *An ambulance of the wrong colour: Health professionals, human rights and ethics in South Africa* (Cape Town: University of Cape Town Press, 1999); Anne Digby, *Diversity and Division in Medicine: Healthcare in South African from the 1800s* (Bern: Peter Lang, 2006).

57 It took over three weeks of working through the collection to come across the first reference to racialized language of any kind.

58 The reasons for the lack of racialized demarcation and segregation evident in the photographs remained opaque. The simplest explanation to me related to another omitted category – that of sex. Coupled with the patient's general

appearance in the photograph, both race and sex may have simply been deemed self-evident. Neither needed to be noted explicitly – seeing the patient may have been seen as enough for the clinical viewer (staff and students imbricated in the habitus of the hospital and South Africa's racialized norms) to confidently deduce these categories. I, however, was unwilling to make these same deductions while entering the photographs into the database.

59 Gesa Grimme, "Systemizing Provenance Research on Objects from Colonial Contexts," *Museum and Society* 18, no.1 (2020): 52-65.

60 Gillian Rose, "Practising photography: an archive, a study, some photographs and a researcher," *Journal of Historical Geography* 26, no. 4 (2000): 561-562.

61 Elizabeth Edwards, "Photographs, Mounts, and the Tactile Archive," *Interdisciplinary Studies in the Long Nineteenth Century* 19 (2014): 14.

62 Julia Bärnighausen, Stefanie Klamm, Franka Schneider and Petra Wodtke, "Photographs on the Move: Formats, Formations, and Transformations in Four Photo Archives," in *Photo-Objects: On the Materiality of Photographs and Photo Archives in the Humanities and Sciences*, eds Julia Bärnig, Costanza Caraffa, Stefanie Klamm, Franka Schneider and Petra Wodtke (Berlin: Max Planck Institute for the History of Science, 2020) 42.

63 Nigel Worden, Elizabeth van Heyningen and Vivian Bickford-Smith, *Cape Town: The Making of a City, An Illustrated Social History* (Cape Town: David Philip Publishers, 1998), 247, 251.

64 Early medical photographs were often also produced by clinicians themselves. Many medical practitioners dabbled in photography, with some arguably serving as key figures in the history of this imaging technology. These include psychiatrist Hugh Welsh Diamond as well as physician and microscopist Richard L Maddox. See Daniel M. Fox, DM & Christopher Lawrence, *Photographing Medicine: Images and Power in Britain and America since 1840* (Connecticut: Greenwood Press, 1988); Mifflin, "Visual Archives in Perspective;" Mienneke te Hennepe, "Depicting Skin: Visual Culture in Nineteenth-Century Medicine" (PhD Thesis: Maastricht University, 2007); Gretchen Worden, *Mütter Museum: Historic Medical Photographs, The College of Physicians of Philadelphia* (New York: Blast Books, 2007); Helmut Gernsheim, *The History of Photography: From the Camera Obscura to the Beginning of the Modern Era* (London: Thames and Hudson, 1969). Indeed, in South Africa, Arthur Bensusan both practiced clinical photography and pub-

lished papers on the role of this medium in medicine on the African continent (SAHO 2013). South Africa's Joseph Denfield too was both a medical doctor and an avid art photographer. Phindezwa Mnyaka, "'Re la Tsoantso' ('Father of the pictures'): Joseph Denfield's photography, 1944-1965," *Kronos* 36 no. 1 (2010): 84-107.

65 Isabel Tang, *Pornography: the secret history of civilization* (London: Chanel 4, 1999).

66 Gernsheim, *The History of Photography*; David M. Frohlich and Risto Sarvas, *From Snapshots to Social Media: The Changing Picture of Domestic Photography* (London: Springer, 2011).

67 A stereographic card consists of two photographs mounted on a single surface. While professional stereographs would usually be made with a specialised camera containing two lenses or a single stereoscopic lens adaptor, the examples in the UCT DoS Collection suggest the less expensive (and less reliable) method of manually shifting the camera to take two photographs of the same subject eye-distance apart.

68 Highly textured paper can be popular for artistic prints, as they provide the illusion of additional depth and are pleasant to the touch when printed in books or as loose, individual objects. As a whole it tends to 'soften' the photograph – a contra-indicative attribute for images meant to objectively relay clinical information without human interference.

69 Jan Hendrik Louw, "Editorial: Charles FM Saint," *SAMJ* 47, no. 8 (1973): 282.

70 Ibid., 282.

71 Cabinet cards are photographs mounted on various standard sized cards that were produced by commercial photographic studios in the 19th and early 20th century. They are often printed in multiples and frequently carry elaborate advertisements for the studio on the back, and are often stamped with the insignia of the studio. Gerald J. Munoff and Mary Lynn Ritzenthaler, "History of Photographic Processes," in *Photographs: Archival Care and Management*, ed. Mary Lynn Ritzenthaler and Diane Vogt-O'Connor (Chicago: Society of American Archivists, 2006).

72 Charles Frederick Morris Saint, "Some Recollections of the Early Days of the Medical School, Cape Town," *South African Medical Journal* 37, no.3 (1963): 49-51, 50.

73 CT Fitz-Simon, "A Historical Review of the Camera's Use in Medicine," *Me-*

dical Historian: The Bulletin of the Liverpool Medical History Society no.2 (1989): 19-26; Kathy McFall. 2000. *A Critical Account of the History of Medical Photography in the United Kingdom.* (Masters) Thesis (University of Wales College of Medicine, 2000).

74 Louw, *In the Shadow*, 173-174.

75 James Rutherford Morison and Charles Frederick Morris Saint, *An Introduction to Surgery* 3rd Edition (Bristol: John Wright & Sons, 1935), vi.

76 Tilli Tansey, "Keeping the Culture Alive: The Laboratory Technician in Mid-Twentieth-Century British Medical Research," *Notes & Records of the Royal Society* 62 (2008): 79.

77 This approach is officially noted in the institutional archive in a 1949 Memorandum of Agreement for technicians that specifically stipulated that "[t]he Technical Assistant shall work under the direction of the Head of the Department or of such members of the staff of the department as the head may from time to time depute, and shall carry out all such instructions as are given to him/her by such person placed in authority over him/her" UCTAA. Personnel Files, Carl Conrad Goosen 4.2.6(358): University of Cape Town Memorandum of Agreement, 13.12.1949.

78 Norman Levy, "History of Medicine: Somerset Hospital: South Africa's Oldest Hospital," *SAMJ* 100, no. 6, (2010): 358.

Jan Hendrik Louw, "The New Somerset Hospital as a Teaching Institution," *SAMJ* 36, no.33, (1962): 674.

79 Michaela Clark, "Designing the Clinic: Racialised Architecture & Cape Town's Old Groote Schuur

Hospital," in *Space and the Hospital* (Turnhout: Brepols, 2025 forthcoming).

80 Louw, *In the Shadow*, 337, 363.

81 James Rutherford Morison and Charles Frederick Morris Saint, *An Introduction to Surgery* 4th Edition (Bristol: John Wright & Sons, 1948), Preface.

82 Lorraine Daston, *Things That Talk* (Brooklyn: Zone Books, 2004), 10.

6

AI-Enhanced Appraisal and Processing of a Hybrid Collection

Tommy Keswick, Research Applications Developer, Caltech Library
Mariella Soprano, Senior Collections and Special Projects Archivist, Caltech Library

The field of archival science is undergoing a remarkable transformation as artificial intelligence (AI) emerges as a potentially powerful tool for processing and managing complex collections. Hybrid collections, which consist of both physical and digital records, present unique challenges that have traditionally been difficult to address using traditional methods. The integration of AI offers an unprecedented opportunity to overcome these challenges, allowing archivists to handle larger volumes of material, streamline workflows, and enhance accessibility for researchers.

At the ICA-SUV 4th Workshop on Scientific Archives, Tommy Keswick, Digital Technologies Development Librarian, and Mariella Soprano, Senior Collections and Special Projects Archivist, presented an innovative approach to managing one of the most significant hybrid collections the Caltech Archives has encountered: The Robert H. Grubbs Papers. Their presentation highlighted a practical application of AI in archival science, offering a case study concerning this technology's potential to revolutionize the field.

This case study focuses on the extensive collection of Dr. Robert H. Grubbs, 2005 Nobel laureate in chemistry whose pioneering work in organic synthesis fundamentally altered the landscape of chemistry and materials science. The processing of his papers offer an ideal opportunity to explore how AI and readily available digitization tools could be applied to complex hybrid collections. The article describes the project's proof-of-concept project as well as initial/early work on its implementation.

The Legacy of Robert H. Grubbs and His Collection

Robert H. Grubbs was awarded the 2005 Nobel Laureate in chemistry for his groundbreaking contributions to the development of the metathesis method in organic synthesis. This innovative chemical process enabled the synthesis of customized molecules, leading to the creation of specialized plastics and pharmaceuticals with far-reaching implications for both industry and medicine. Over the course of his career, Grubbs authored over 200 patents, established several companies, and worked as a consultant for numerous organizations.

The collection is a testament to Grubbs' prolific career. Comprising more than 100 bankers' boxes and containing a vast array of digital files, the materials include correspondence, patents, research data, administrative records, and personal documents. (Figure 1)The digital records consist of data retrieved from three different computers, dozens of floppy disks, and email accounts spanning multiple platforms.

To complicate matters further, the digital records of Grubbs' longtime administrative assistant were also included in the collection. While the assistant had organized the physical materials into filing cabinets and

Figure 1

labeled the drawers by broad categories, there was no overarching system to integrate these with the digital files. This lack of a unified structure presented a significant barrier to access, as researchers would need to navigate both physical and digital domains to locate relevant information.

Initial Challenges and Strategic Objectives

The Grubbs collection poses two significant challenges to the archivists tasked with its processing. The sheer size of the collection alone is daunting, while its hybrid nature required a comprehensive strategy that could effectively bridge the gap between physical and digital records.

After consulting with library leadership and colleagues with technical expertise, we outlined a series of strategic objectives to guide our work:

1. Reducing Physical Volume: One of the first priorities is to identify and discard duplicate records, particularly printed copies of digital documents. By reducing redundancy, the team aims to free up valuable storage space and streamline the organization of the collection.

2. Developing an Inventory: Another key priority is to provide researchers with a detailed content list of the papers included in the collection. The team aims to create a spreadsheet inventory of the titles of the physical archival boxes and, within those, the titles of the physical folders. The box and folder titles will be based on the labels applied by Grubbs' assistant, with the box titles reflecting the labels on the filing cabinet drawers, which represent the primary intellectual subdivisions. The goal is to offer researchers a comprehensive roadmap to the physical materials. Additionally, the inventory will include the creation dates of the documents.

3. Facilitating Remote Access: To enhance remote access to the collection, the team aims to generate preview images of key documents, which will enable researchers to review materials before visiting the archives. The intention is to provide a quick view of the content without attempting to provide high-quality images. It is expected that researchers will request high-quality images if needed.

4. Exploring AI Applications: To enhance the organization and processing of the collection, the team will continue to experiment with AI-driven techniques, including natural language processing (NLP) and machine learning.

5. Creating a Unified Finding Aid: A key long-term goal is to develop a cohesive finding aid that integrates both physical and digital materials, providing researchers with a seamless and comprehensive guide to the collection.

Automated Image Capture & Organization

To analyze the contents of the physical collection, we needed a way to capture the text in the documents. Without the funding for a full digitization project, we decided photographing the documents as they were being evaluated for retention would be an effective strategy. We knew that we could use digital photos of documents in an asynchronous OCR workflow to analyze the contents of the physical files.

An early primary concern with this plan was how to avoid losing the hierarchy of box/folder/document in a "camera roll" on a mobile device. We needed to label the photographs with a meaningful filename and ideally mirror the physical hierarchy with a digital directory structure for the image files. The basic camera applications on mobile phones do not allow for this kind of organization without manual intervention on the device after taking each photograph.

To address the hierarchy needs, we have developed a set of custom workflows with the Apple Shortcuts application, which is a no-code automation tool available on iOS, iPadOS, and macOS. We subsequently dubbed our custom application *Popstar: **Pho**ne-based **p**rocessing **s**of**t**ware for **ar**chives.*

The Popstar application enables the organization of digital images in a directory hierarchy with as little human interaction as possible. Shortcuts has Actions available for taking photographs, extracting text from images, creating directories in iCloud Drive, and capturing user text input, among many others. Popstar does rely on a single third-party application called Toolbox Pro for Shortcuts in order to allow the different custom workflows to pass data about the hierarchy between each step of the Popstar application.

Popstar includes several key components that help address our needs:

- **Directory Creation & File Organization**: Popstar creates directories in iCloud Drive to maintain the hierarchy. It takes a user-entered project name to create a top-level directory, and then creates directories for boxes and their folders based on captured text from labels that we place under the camera. Files are sequentially named based on their folder labels and an auto-incrementing number.
- **Image Capture**: Photos are taken that will serve as a source for OCR text and as preview images that can allow researchers to get a sense of the collection.
- **Streamlined Integration**: The software integrates into our physical workflow to bolster efficiency while capturing a digital representation of the collection contents.

By automating tasks of scanning and organizing digital images, Popstar makes it possible to process materials more quickly and affordably than traditional methods would have allowed.

A New Processing Workflow

Much consideration was given to developing an efficient physical workflow that takes place in a small processing room. We had to accommodate a photographing station, a label-making station, and a weeding station, all among the scores of unprocessed and preliminarily processed boxes.

We settled on a portable standing desk on casters with a large ring light and phone clamp positioned above the desktop for the photographing station. (Figure 2) Tapping on the phone to take photographs while it is clamped within the ring light would be too disruptive to the somewhat-delicate positioning of all the devices so we connected an external Bluetooth keypad to the phone to engage the shutter button for taking photographs. (Figure 3) Additional keys on the keypad function as triggers for the custom Popstar *Shortcuts* application to create the directory hierarchy within iCloud Drive on the iPhone. In this way, our photographer does not have to touch the phone at all, but simply places labels or pages on the desktop under the camera and taps the appropriate key on the keypad to create directories and take photos of the documents.

The full in-room workflow is as follows:

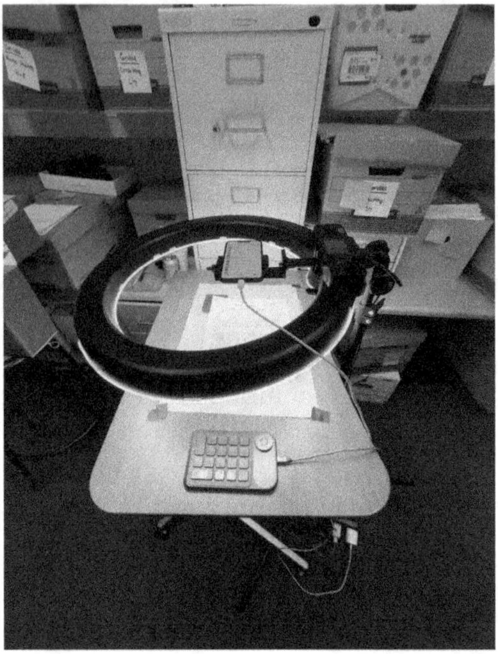

Figure 2

Figure 3

- A box label is positioned under the camera and the keypad button programmed for "create a new box directory" is tapped.
- A folder is positioned (with or without a new printed label attached, depending on the legibility of the original label) under the camera and the "create a new folder directory" button is tapped.
- The folder itself is photographed using the "take a photo" button on the keypad.
- The first page in the folder is moved under the camera and the "take a photo" button is tapped.
- Page photographing continues until the end of the folder.
- The next folder is positioned, a directory is created, and the photographs are taken.
- The next box would be started by beginning the steps again.

For some sections of the collection we have to modify this general workflow. For example the boxes containing publications by Grubbs have folder labels with the complete title of articles, which can run to dozens of words. There are technical limitations to the length of directory and file names, and we knew we would run up against these if we used the original, too-lengthy titles. We adapted the workflow to generate a new shorter, unique label for a publication folder and use that for directory and file names. We then took an additional photograph of the original label to capture the full title of the article.

Adapting Traditional Appraisal and Processing Methods

The team is taking care to adapt traditional archival methods to meet the unique requirements of the Grubbs collection. Our aim is to strike a careful balance between accomplishing the time-tested work of archival appraisal and processing and making necessary adjustments to accommodate our additional goals.

Retention Decisions: Not all materials in the collection are deemed worth preserving. For example, travel files such as airline tickets and hotel bookings will be excluded from the final archive, as they hold little long-term research value.

Maintaining Original Order: The physical arrangement of the collection, which had been established by Grubbs' assistant, serves as the basis for the finding aid. This preserves the context in which the materials were originally used, while also providing a logical framework for researchers.

Relabeling and Preservation: Many of the folders were housed in hanging file folders, which are not ideal for long-term preservation. These are replaced with archival-quality folders, and the contents are relabeled to ensure consistency and clarity.

Selective Digitization: To optimize efficiency, the team is adopting a selective approach to digitization. For multi-page documents, only the first page is to be photographed—unless it lacks intellectual significance, in which case the next relevant page is selected. Multiple-page documents that have not been photographed are to be marked with an orange paperclip, indicating that additional pages exist in the original.

Photographic Documentation: To establish a clear digital structure, the team is to photograph the box labels and folder labels, which are then to be used to create a consistent naming convention for the digital files.

Experimental Analysis

Once some initial photographing took place, we began experimenting with tools for analyzing the contents of the documents. We extracted the text from the photographed documents using a different Apple Shortcuts routine, mostly because it was readily available on macOS and because it was interesting to familiarize ourselves with another use for the application. Any OCR application would have been suitable for this task.

We ran some scripts that were developed in-house that use common Natural Language Processing (NLP) algorithms to compare the photographed documents to the entire corpus of born-digital files from Grubbs' computers and email accounts. We also used some Named Entity Recognition (NER) algorithms to find names of interest in the documents.

An additional, not initially planned, experiment that we ran was to provide the ChatGPT API with some of the text extracted from the photographed documents to see what it would tell us. After observing initial results, we iteratively adjusted the prompts to refine the types of information that it returned.

The integration of machine learning tools into the archival workflow presents one of the most exciting and forward-looking aspects of the project. By leveraging these types of applications, the team will be able to automate many of the tasks traditionally performed manually, such as identifying duplicate records, extracting metadata, and analyzing thematic content.

Algorithmic Methods and Predicted Outcomes

Document Similarity Analysis: Algorithms and large language models can be used to identify duplicate records within the collection. This will be particularly useful for eliminating printed copies of digital documents, reducing redundancy, and streamlining the organization of the materials.

Named Entity Recognition (NER): NER models should prove useful in extracting names, dates, and other key entities from the documents. This information will then be used to create a comprehensive index, making it easier for researchers to locate relevant materials.

Topic Modeling: Natural language processing tools can be employed to group documents by thematic content. This will make it possible to identify trends and patterns within the collection, providing valuable context for researchers.

Confidentiality Detection: Advanced search tools will also be used to identify sensitive materials, such as documents marked "confidential." These materials will be flagged for restricted access, ensuring that privacy and confidentiality are maintained.

Potential of ChatGPT: The team is experimenting with the ChatGPT API to summarize documents and extract key metadata. While this approach shows promise, its scalability for large and multi-faceted collections remains an open question.

Preliminary Results and Lessons Learned

The early results of this approach are promising. By applying AI techniques and leveraging tools like Popstar, the team successfully can streamline many aspects of the processing workflow. Key takeaways from the initial exploratory phase of the project include:

1. Improved Efficiency: Algorithmic tools should significantly reduce the time required to identify duplicates, extract metadata, and organize materials, making it possible to process a larger portion of the collection in a shorter time frame.

2. Enhanced Accessibility: The combination of a detailed inventory, preview images, and thematic categorization can make the collection more accessible to researchers, including those accessing it remotely. This should be especially beneficial for researchers who might otherwise be facing geographic or time constraints.

3. Preservation of Original Order: By respecting the organization established by Grubbs' assistant, the team is maintaining the integrity of the collection while improving its usability.

4. Scalability Challenges: While AI tools like ChatGPT and NER models show potential, their application to large-scale collections raises questions about scalability. For example, the computational resources required to analyze massive datasets and the limitations of existing AI models pose challenges that will need to be addressed in future projects.

5. Interdisciplinary Collaboration: The success of the proof-of-concept phase of the project underscores the importance of collaboration between archivists and technologists. Each is contributing unique insights that enhance the overall strategy.

6. Workflow Adaptability: Ensuring that the software and workflows used to complement the traditional appraisal and processing methods can be adapted to unexpected situations will be crucial in a large hybrid collection. Different sections of material may require flexibility in how they are organized and photographed.

Further Questions and Considerations

Early work on the project has also raised several important questions that highlight the broader implications of using large language models in archival science:

1. Policy Development: What policies need to be established to guide the use of AI in archives? Issues such as data privacy, intellectual property, and ethical considerations must be addressed to ensure that AI is used responsibly.

2. The Role of Archivists: How will AI reshape the role of archivists? While AI can automate many routine tasks, archivists will still play a critical role in decision-making, interpretation, and ensuring the contextual integrity of collections.

3. Hybrid Finding Aids: How can AI be leveraged to create finding aids that seamlessly integrate physical and digital materials? The development of such tools will require further innovation and experimentation.

4. Long-Term Preservation of AI-Generated Metadata: As archives increasingly rely on AI to generate metadata, how can we ensure that this metadata remains accessible and interpretable over the long term?

5. Training and Resource Allocation: What training and resources will be required to equip archivists with the skills needed to use AI effectively? Institutions will need to invest in professional development and technological infrastructure to fully realize the potential of AI.

Conclusion

Our work with the Robert H. Grubbs Papers represents a persuasive and promising example of how AI and advanced technology can be integrated into archival processing workflows. By combining traditional archival methods with cutting-edge tools like Popstar and AI-driven analysis, the team is able to develop, implement, and refine new strategies for dealing with the challenges posed by this large and complex hybrid collection.

This integrated interdisciplinary approach to the Grubbs processing project serves as a possible model for the future of archival collections work at Caltech and beyond, demonstrating the potential of AI to enhance efficiency, accessibility, and preservation. However, it also underscores the need for ongoing innovation, collaboration, and critical reflection. As the field continues to evolve, archivists and developers will need to navigate new challenges, develop robust policies, and adapt their roles to fully harness the power of AI.

Ultimately, the initial proof-of-concept stage of the Grubbs project offers a vision of what is possible when technology and archival expertise are brought together. This work will not only enhance access to the legacy of a Nobel laureate but also points the way to a new era in archival science. Certainly, more work is required to realize this vision. Continuing to experiment with new tools as they become available will play an important role in keeping archival collections relevant as well as available to wider audiences.

Acknowledgments

We are deeply grateful to the Grubbs family for their generous donation of the Robert H. Grubbs papers to the Caltech Archives.

This project would not have been possible, nor could it have progressed, without the unwavering support of the Caltech Library and Archives.

The initial software prototype of Popstar was created and designed by Mike Hucka, formerly Research Applications Developer at the Caltech Library, now Staff Software Engineer at Google Quantum AI. Stephen Davison, Head of Digital Library Development, has been instrumental in supporting the project and continues to assist with software development. Ian Roberts, Systems Administrator and Manager of the Library Technology Hub, plays a key part in acquiring and setting up many of the physical and digital components of the workspace and workflow.

We extend our gratitude to Richard Thai, Digital Archivist, for his critical role in transferring and organizing the Grubbs emails and digital files. We thank Heidi Aspaturian for her editorial work. The steady progress of this project has been made possible in large part due to the dedicated efforts of our volunteers: Margaret Cribbs and Kate Jackson.

Finally, we sincerely thank the Caltech Division of Chemistry and Chemical Engineering for providing a workspace within their physical facilities and for assisting with any related facilities issues.

7

Reappraising *Appraising the Records of Modern Science and Technology*: Helping Archivists Document Twenty-First-Century Science

Bethany G. Anderson, Polina E. Ilieva, and Jordon Steele

Abstract

In 1985, Joan K. Haas, Helen Willa Samuels, and Barbara Trippel Simmons published *Appraising the Records of Modern Science and Technology: A Guide*, providing essential guidelines for archivists managing the increasing records generated by postwar science. The *Guide* built on prior efforts to develop appraisal guidelines for the extensive records of twentieth-century science, encompassing personal and professional activities, science administration, and research across academia and industry. It highlighted the importance of understanding scientific functions and activities, contributing to broader discussions on functional analysis and documentation strategy.

The scientific landscape has notably evolved since 1985, with the advent of "big data," open science, data management, citizen science, and scientific exchanges on social media, posing new archival challenges. In 2022, a research project was launched to reassess the *Guide* in light of these developments. Archivists from a research center, a health sciences university, and a science-intensive land-grant university conducted a gap analysis of the *Guide* and interviewed scientists, engineers, and medical professionals at their institutions (University of Illinois Urbana-Champaign, Johns Hopkins Applied Physics Laboratory, and University of California, San Francisco). They compared these findings to refine archival practices for contemporary science, addressing the new challenges and directions for the twenty-first century.

Introduction

In 1985, Joan K. Haas, Helen Willa Samuels, and Barbara Trippel Simmons published authoritative guidelines for archivists working with science and technology records. *Appraising the Records of Science and Technology: A Guide*[1] was at the time a much-needed volume for archivists grappling with the exponential rate at which postwar science generated records of its activities. The *Guide* built on decades of work by archivist to develop appraisal guidelines for twentieth-century science's voluminous creation of records, including the Joint Committee on Archives of Science and Technology's (JCAST) final report in 1983 on managing scientific records. However, the *Guide* sought to be broader in scope by taking into account personal and professional activities, the administration of science, and research and development in both academia and industry. The guidelines also sought to instill in archivists the importance of identifying and giving critical thought to the functions and activities resulting from scientific research and processes, linking it to broader theoretical appraisal discussions about functional analysis and documentation strategy associated with Samuels's previous work.[2]

For decades the *Guide* has been a mainstay for archivists, widely read and known for its clear and comprehensive analysis and advice. However, the scientific enterprise has evolved in significant ways since 1985, most notably in the areas of big data, the open science movement, data management, citizen science, and scientific exchange on social media. These changes have raised new questions for archivists about what contemporary records of science should be collected and preserved in perpetuity.

Research project overview

In 2022, the authors of this article launched a research project to reexamine the *Guide* in light of these developments. The research team represents three distinct types of institutions that acquire, preserve, and make accessible records of science, technology, and engineering: a federally-funded research laboratory (Johns Hopkins Applied Physics Laboratory), a health sciences university (University of California, San Francisco), and a science-intensive land-grant university (University of Illinois Urbana-Champaign).

The questions the research project seeks to answer include:

- Is "modern" science and technology practice as of 1985 still modern?
- How have computers and the Internet changed science and its resulting records?
- How have open science and open data movements changed scientific practice?
- Should we rethink what constitutes "scientific archives"? (e.g., include specimens, software, organic tissue material, etc.?)
- Is the *Guide* inclusive of all institutions "doing" science and managing scientific records?
- Do Haas et al. address diversity, equity, inclusion, and accessibility (DEIA) in science?
- What about access to and use of the contemporary scientific record?

The research project comprises the following activities:

- Conduct a gap analysis of the *Guide* to identify new directions and ways to build on the manual to reflect contemporary science
- Interview scientists, engineers, medical professionals, and administrators from the research team's respective institutions to gain deeper insight into what scientific research looks like in the contemporary era and what records are created in support of that activity
- Compare findings from the interviews with the gap analysis
- Recommend updates to the *Guide*'s conclusions that assist archivists and curators tasked with documenting twenty-first century science, technology, and engineering

Research to date

Phase 1: Gap Analysis

In early 2022, the research team examined the *Guide* and created a series of tables outlining the major functions, associated activities, and related records identified in the manual. Drawing on its collective experience with twenty-first century scientific archives, the team then evaluated the extent to which the *Guide*'s fundings matched contemporary science, recording in the spreadsheet instances where contemporary scientific records may take a different form than those described in the *Guide* or may be new types of records created since Haas, et al., published the *Guide*. For example, memoranda in paper format today have been replaced by emails and formal memoranda as digital documents (e.g., PDFs). Many of the functions identified in the *Guide* matched the research team's perceptions of functions governing contemporary science. In many cases, however, technological advances since 1985 (namely the normalization of personal computing and the development of the Internet) changed both the form of the records reflecting these functions and recordkeeping behaviors practiced by stakeholders to manage the records. Email has accelerated and increased the amount of records documenting the early stages of the scientific processes. File-sharing platforms like Box and Google Drive have made collaboration and sharing faster and easier, but they have also decentralized recordkeeping, resulting in a loss of control not seen in the pre-digital era. And social media services have weakened communication barriers among scientists and between scientists and the general public but have also raised questions about what is the "record copy" of this content and who owns it given shifting terms of service for many of these platforms. The research team shared these and other initial findings at the Society of American Archivists's Research Forum in August 2022.

Phase 2: Interviews

After completing the gap analysis, the research team sought to test what constituted an educated hypothesis about the state of contemporary science by conducting interviews with people principally involved in the scientific enterprise. To ensure consistency in data-gathering, the team adopted a single set of questions to ask interview subjects at each institution. (See Figure 1) Each team member identified 4-5 people at their institutions to interview. The team aimed to interview scientists, engineers, and administrators across a diversity of scientific fields and different career stages with the idea to capture a representative cross-section of the scientific enterprise. In some cases, recordkeeping is a collaborative activity which meant that some interviews were conducted with multiple individuals. In other cases, we ended up interviewing different individuals than those we initially contacted because the lab/research center/department had a position devoted to recordkeeping and management (at least in part). The professional profiles of the interview subjects selected were as follows:

Johns Hopkins Applied Physics Laboratory:

- One late-career scientist in a leadership role, specializing in cyber intelligence systems
- One late-career scientist in a leadership role, specializing in physics
- One late-career engineer considered a technical expert, specializing in optical engineering
- One early career mathematician, specializing in data science
- One program manager with administrative responsibilities

University of Illinois Urbana-Champaign:

- One advanced career engineering professor and the manager of their lab
- Two early-mid career archaeologists
- One mid-career waterfowl biology research scientist

University of California, San Francisco:

- One late-career professor of pathology and director of a clinical lab
- One mid-career clinical professor of cytogenetics and assistant director of a clinical lab
- One early-career assistant professor of biomedical engineering and PI of a research lab

- One mid-career professor of microbiology and immunology and PI of a research lab
- One postdoctoral fellow in biochemical and molecular nutrition and researcher in a lab

All three authors inquired with their Institutional Review Bords (IRB) about the need for their oversight to conduct the interviews. JHU/APL did not require IRB approval. UCSF and University of Illinois determined this research to be exempt since it does not involve human subjects as defined by the federal regulations. Both authors from these institutions received formal "not human subjects" determination letters from the IRB citing the regulations to be shared with interviewees.

When contacting prospective interviewees, the research team used the following explanation in an effort to be explicit about what was meant by "the scientific record" to non-archivists:

> The goal of our research project is to better understand the ways in which new technologies and scholarly communications have impacted scientific records in order to assess how current archival practices need to evolve to attend to emerging needs. Different professions have different understandings of archives and records, so for the purposes of this research project, we define an archival record as "Materials created or received by a person, family, or organization, public or private, in the conduct of their affairs that are preserved because of the enduring value contained in the information they contain or as evidence of the functions and responsibilities of their creator."[3] For scientific records specifically, this could include email correspondence, drafts of publications, grant proposals and reports, research notes and notebooks, datasets, and social media content. We are seeking to understand the spectrum of documentation created, how staff use and share these materials, and staff's recordkeeping practices for collaborative research projects.

Beginning in 2023, the team recorded semi-structured interviews over

Zoom or in person, each lasting about an hour, to learn more about the subjects' scientific work, what records they created in support of this work, and what recordkeeping practices they undertook to manage these records. The team presented on its initial findings at the August 2023 meeting of the Society of American Archivists's Research Forum and provided an update at the Fourth Workshop on Scientific Archives in 2024. Initial analysis of the interviews conducted to date include the following findings:

Johns Hopkins Applied Physics Laboratory:

- Primary deliverables for interviewees mainly include reports and PowerPoints
- Paper notebooks are still used (at the bench and in classified environments)
- Not heavy consumers of social media
- Universal concern about decentralized/absence of recordkeeping in digital age

University of Illinois:

- Primary deliverables for interviewees include publications, reports, and data
- Paper notebooks are still used (in the field and in the lab, but some digitize and/or transcribe them for broader sharing with colleagues)
- Use of multiple platforms and systems for collaboration and communication (e.g., Slack, Dropbox, Google Drive, and Mural)
- Recordkeeping practices informed in some case by federal and state funding requirements but more often driven by ongoing research needs and (re)use
- Social media is not frequently used
- Organization is important for those working in lab settings with other colleagues to facilitate sharing, use, and longitudinal projects
- Some researchers have recordkeeping functions built into their job responsibilities, or employ staff members for whom recordkeeping and management is central aspect of their position

University of California, San Francisco:

- Primary deliverables for interviewees include conference presentations, book chapters, data, peer-reviewed journal publications, software, cell lines, images, and bacteria
- Physical notebooks are still a preferred method. For wet lab research only physical lab notebooks are used because researchers wear gloves and deal with chemical agents. Summaries and data are then recorded in digital formats. Only one lab requires use of E-lab notebooks
- Researchers use multiple platforms and systems for collaboration and communication (e.g., Slack, Box, shared institutional servers, Zoom, email, Microsoft OneDrive, Google Drive, GitHub, LinkedIn, listservs)
- Recordkeeping practices are informed by institutional, field, federal, and state funding requirements and data management and sharing policies
- Social media is not used for sharing own results, occasionally used for learning about others' work and methods (however, since these are not peer-reviewed scientists need to check peer-reviewed publications before reusing these methods)
- All labs have websites that provide details about research projects, lab members, publications, awards, videos of experiments
- Majority considers their laptop computers their primary archives
- Many researchers are experimenting with AI. UCSF developed a secure generative artificial intelligence (AI) platform – Versa, that consists of two products – Versa Chat and Versa API. UCSF best practice is to disclose AI use.
- The research team noticed several trends across each institution. One notable takeaway is that contrary to initial assumptions, social media is not used and has little role in professional communication and in disseminating research findings. Additionally, paper records (especially lab or field notebooks) continue to be one of the main outputs of scientific research and administration. However, many

of these paper records are converted into a digital format to facilitate access across distributed working environments (especially in cases where some individuals work remotely part of the time). To that end, the research team also found that cloud-based sharing platforms (such as Box, Google Drive, and Dropbox) are universally used. Based on the team's assessment so far, it appears that the creation of the records has not changed significantly, but the ways in which they are accessed have.

Conclusion and next steps

The research team intends to complete interviews in 2025. After this, the team will perform a qualitative data analysis using coding techniques. A list of codes that will be commonly used to tag passages in across the interviews has been created. Codes include activity-based tags (creation of records, accessing records, keeping current, etc.), process-oriented tags (file-naming, retention mandate, etc.), and tags identifying types of records (See Figure 2). For code analysis, Taguette[4] was selected as the qualitative data analysis tool given its flexibility and ability to be used across different institutions as well as being free and open access. It can also be installed locally on a device or its cloud-based platform can be used, which enables institutions with different security requirements to access and use it.

Once coding is completed, the research team will analyze the tags to draw conclusions about what themes emerge from the interviews. The team will close its study by comparing these themes with the *Guide* to inform recommendations for additions or updates to the manual's recommendations to reflect scientific practice in the twenty-first century. The research team aims to complete the project by end of 2025. Our intention is to publish our results in a peer reviewed journal. Given that this project was based at institutions in the United States, and the issues that impact scientific archives are global, the research team hopes to expand the project and undertake a second phase that examines these issues internationally in collaboration with archival partners around the world.

Figure 1. List of interview questions

- How would you describe research records and administrative records in scientific enterprise?
- What materials do you create and reference that help you do your work?
- What are the scientific outputs/deliverables of your work?
- What recordkeeping guidelines are you required to follow?
- What recordkeeping guidelines do you follow?
- In your unit [department/institute/lab/etc.] who has a recordkeeping responsibility?
- What records (scientific and administrative) do you keep and why?
- Where and how you preserve these records?
- Where do you keep your physical records? [if meeting in person]
- Where do you keep your digital records?
- Do you work collaboratively, and if so, how do you collaborate (email, chat, collaborative virtual working spaces, etc.)
- Do you share your own records or use the records created by colleagues? Have you ever annotated or added to the records created by other colleagues (or had them annotate/add to records you created)?
- Who is responsible for record keeping on a multi-institutional collaborative project? Does each institution keep records for research they produce? How and where are the records kept and shared after the conclusion of a collaborative project?
- How do you share your research findings?

Figure 2. Tags

- Access barriers
- Access restrictions
- Administrative records
- Analog records
- Audiovisual records
- Classified/unclassified records
- Collaboration
- Collaborative platforms
- Destruction of records
- Digital records
- Email
- Equipment use
- Federal/state mandates
- Field/lab notebooks
- File-naming
- Grant records
- Legacy media
- Objects
- Organizational schema
- Patents
- Presentations
- Protocols
- Publications
- Recordkeeping in the "field"
- Recordkeeping in the "lab"
- Reformatting data
- Reports
- Research data
- Research records
- Research protocols
- Retention
- Social media
- Software
- Specimens
- Sponsor/funder relationship
- Storage
- Use of records
- Websites

Endnotes

1 Haas, Joan K., Helen Willa Samuels, and Barbara Trippel Simmons. *Appraising the Records of Modern Science and Technology : A Guide.* ([Cambridge] Chicago: Massachusetts Institute of Technology, 1985); Distributed by Society of American Archivists.

2 For more information about her previous writings on appraisal, see Helen Willa Samuels, Varsity Letters: Documenting Colleges and Universities (Scarecrow Press, 1998); and Helen Samuels, "Who Controls the Past," The American Archivist 49, no. 2 (April 1986): 109-124. doi: https://doi.org/10.17723/aarc.49.2.t76m2130txw40746.

3 Dictionary of Archives Terminology, s.v. "archival record," accessed May 15, 2025, https://dictionary.archivists.org/entry/archival-record.html.

4 https://www.taguette.org/.

The Appraisal of Scientific Documentation/Information as a Methodological Operation: Application Criteria and Parameters

Fernanda Ribeiro
fribeiro@letras.up.pt
ORCID 0000-0002-5641-9199
University of Porto, Faculty of Arts and Humanities, Portugal
CITCEM - UIDB/04059/2020
DOI: https://doi.org/10.54499/UIDB/04059/2020

Armando Malheiro da Silva
malheiro@letras.up.pt
ORCID 0000-0003-0491-3758
University of Porto, Faculty of Arts and Humanities, Portugal
CITCEM - UIDB/04059/2020
DOI: https://doi.org/10.54499/UIDB/04059/2020

Abstract

Evaluation in the context of information services has been carried out essentially in three distinct areas: the evaluation of services, the evaluation of information retrieval and the evaluation (appraisal) of information flow. This last aspect, especially applied to archives, aims to decide the destination of information after a few years of current use, seeking to make eliminations that considerably free up the information storage space. In this paper, evaluation is approached not as a practical or merely technical procedure, but as a methodological operation applicable to information in any production and use context, within the framework of information science, an area in which we integrate archivistics as an applied discipline. Appraisal does not, therefore, have an end in itself, disconnected from a broader method that associates the comprehensive or scientific aspect with the applicational or technical one, as it is appropriate in an applied social science such as Information Science.

Starting from the appraisal model, developed at the University of Porto, Portugal, criteria and parameters are set out to, in an objective way, moving towards practical application, taking into account the life cycle of information, the renewal and obsolescence of knowledge and the importance of memory for the long-term preservation.

Introduction

The problematics of information evaluation has been analyzed by the authors at various times and has already given rise to some works, which form the basis for the synthesis developed in this text (Ribeiro and Silva 2004; Silva and Ribeiro 2014).

Looking at evaluation procedures in the field of information science (IS), it can be observed that there has been a dispersion of criteria and parameters associated with practical procedures, which are 'closed' in diverse approaches to information and information services. This dispersion, which includes several facets that can be evaluated, can nevertheless be systematized and grouped into three major areas of specific application, namely:

1. The evaluation of **information services' performance**, an approach that has been accumulated for decades and is closely related to the quality of Libraries, Documentation Centres and Information Services functioning. This aspect of evaluation is based on theoretical and practical references that are not specific to IS, but rather interdisciplinary in nature. In fact, the theoretical foundations or guiding principles for the practice of evaluating information services should be sought in administration and management sciences. Evaluating the performance of information services is, in essence, no different from evaluating other services, despite their own specificities which need to be recognized, but which also exist in other types of services, be they in the financial, commercial, industrial or cultural areas.

2. The evaluation of the **effectiveness and efficiency of information retrieval**, provided by inventories, catalogues, databases, in short, all the information access tools. In this dimension, the literature and practical work produced emphasize the criteria and models designed to analyze and evaluate the organization and representation of information, that is, the quality of its technical processing measured according to the specific users' needs. The dimension of the specificity and effectiveness of representation/ search tools is a facet of the evaluation perspective referred to in 1, since the performance of information services involves, among many other things, evaluating the tools that make it possible to retrieve information. But in practice, this aspect of information systems performance has been studied as a single variable, quite autonomous and subject to very specific evaluation criteria and parameters.

3. Evaluation of the **information flow** of any kind of entity/ organization, produced/received and accumulated organically, that is, in the course and as result of an activity, with a view to preserving continuous information records and eliminating what does not objectively need to remain in present and future memory, thereby thickening and blocking it. This third area is the one we are most interested in addressing here because we believe it is, from a theoretical-methodological point of view, a priority within the scope of research activity in IS. Evaluation work and studies in this area have mainly been carried out in the field of archives, but we believe that the evaluation of information flow should be extended to all contexts that generate, produce and accumulate information, and it is in this sense that we present a new theoretical-practical evaluation model.

Theoretical-Methodological Framework

It is not the purpose of this text to provide a detailed explanation of the theoretical-epistemological foundations supporting the scientific-informational paradigm within which IS is situated, according to how it is conceived and practiced at the University of Porto (Silva and Ribeiro 2002). However, in broad terms, it is important to highlight the essential pillars that support this perspective: the assumption of *Information*, and not the document, as the object of work and study; the adoption of the *quadripolar research method*, developed within the framework of the social sciences; and the preference for *system theory* as an interpretative tool and reference for any scientific (pure or applied) research.

Starting from the following definition, we must value certain aspects of the information itself that have been disregarded or even ignored in a documentary approach:

> *Information* is a structured set of codified mental and emotional representations (signs and symbols), modelled with/by social interaction, and capable of being recorded on any material medium (paper, film, magnetic tape, compact disc, etc.) and, therefore, communicated in an asynchronous and multidirectional way (Silva and Ribeiro 2002, 37).

Considering information as a human phenomenon whose genesis occurs in the brain, closely linked to the individual action in a socially contextualized manner, the study of this phenomenon must lead us to cognitive and sociological approaches, in clear interdisciplinary alignment with IS, a perspective that has been largely absent from the so-called documentary sciences.

Research on the conditions that generate information require valuing the organic context associated with it, which calls for a deep redefinition of theoretical perspectives and technical procedures. On the other hand, the knowledge of the conditions under which information is sought and used, while also requiring interdisciplinary studies (behavioral, psychological, and sociological), invokes the need for profound revisions in library and archival methods related to the treatment of information, which we believe can only be re-examined within IS theoretical framework.

However, the definition of information alone is insufficient for characterizing it as an object of study, and it must be complemented by a statement of its inherent properties/attributes, namely:

- *structured by an action* (human and social) – the individual or societal act structurally establishes and models information;
- *integrated dynamically* – the informational act is involved with, and results from, conditions and circumstances both internal and external to that action;

- *has potentiality* – a statement (to a greater or lesser extent) of the act which founded and modelled the information is possible;
- *quantifiable* – linguistic, numeric or graphic codification is capable of quantification;
- *reproducible* – information can be reproduced without limit, enabling, therefore, its subsequent recording/memorization;
- *transmissible* – informational (re)production is potentially transmissible or communicable.

If the proposed definition of information and its properties have consequences at various levels, the adoption of the 'quadripolar method', designed by P. De Bruyne, J. Herman and M. De Schoutheete as a qualitative research dynamics is no less significant and has implications of various kinds (De Bruyne, Herman and De Schoutheete 1974). The following diagram (Figure 1) shows the interaction of the four poles of the methodological device. The application of the quadripolar method illustrated in Figure 1 shows the interaction and simultaneity of the poles, which is typical of exploratory work on human and social phenomena and problems.

The research dynamics results from an interaction between four poles – *epistemological, theoretical, technical, and morphological* – allowing for a continuous projection of interpretative paradigms, theories, and models in the operationalization of the research and in the presentation of its results.

The ***epistemological pole*** – the scientific community of information professionals, their schools, institutes, working places, with their own political, ideological and cultural references – operates the permanent construction of the scientific object and the definition of the boundaries of the research problems. The discursive parameters are constantly reformulated, as are the paradigms and scientific criteria (objectivity, reliability and evaluation) which guide the whole research process. Empirical procedures and archival knowledge gradually substantiate this pole, which is by no means static but, on the contrary, must be subject to periodic reflection on the occurrence, or otherwise, of epistemological continuity or gaps.

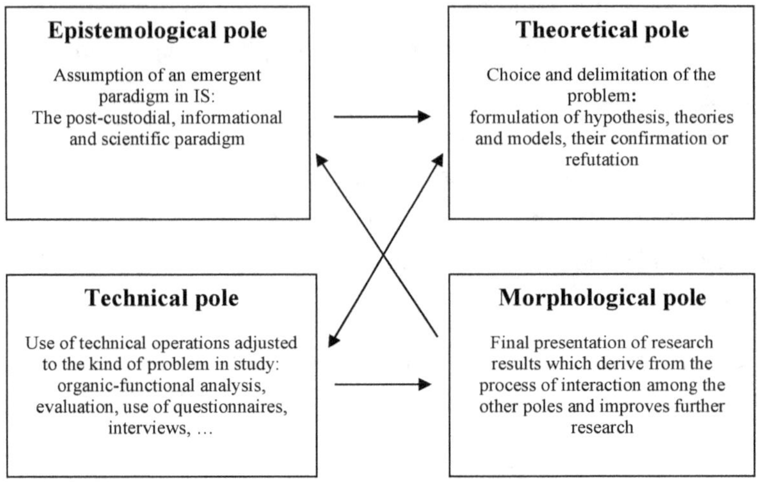

Figure 1. Schematic representation of the Quadripolar Method and the interaction of the four poles

The ***theoretical pole*** operates the rationality of the subject (who knows and approaches) over the object, as well as the postulation of laws, the formulation of hypotheses, theories and operational concepts and the consequent validation or refutation of the "theoretical context" elaborated. This pole supports the technical and instrumental component and gives meaning to the presentation of results, which substantiate the morphological pole. Although there are different theories and models that support ways of thinking about the informational phenomenon/process, we express our preference for System Theory, originated from the studies of Ludwig von Bertalanffy in the 1920s, as it integrates a holistic vision and fits well within the complex and diffuse world of information, as evidenced by several practical applications of their theoretical approach. *General System Theory* (Bertalanffy 1976) has been widely applied and developed to the point where it is now more appropriate to speak of "systems thinking," which is primarily a way of conceiving reality, whatever it may be (Mella 1997).

On the ***technical pole***, the contact with objectified reality is operated through instrumental application, thus verifying the validation capacity of the methodological mechanism. This pole includes three major operations (Lessard-Hébert, Goyette and Boutin 1994):

1. *Direct and indirect observation* (of cases or variables) – comprehensive collection of historical-institutional, legal, and regulatory elements, and the description of the nature and internal functioning of any information system (case/cases) or any supra/inter-case subject (variable/variables) through various techniques such as surveys, interviews, participant observation, and the subsequent elaboration of record forms.
2. *Experimentation* – in a controlled, closed, and artificial field (as opposed to the real, open, and uncontrolled field of case studies), which is enhanced in a laboratory or computer simulation setting, a clear problem formulation, the description of analysis techniques, and the presentation of the instruments, equipment, and scientific means used are essential. If based on sampling, it relies on statistical analysis according to logic hypothesis testing.
3. *Retrospective and prospective analysis/evaluation* – a) general statement: submit the results of observation and/or experimentation to a rigorous examination allowing for generalizations and the establishment of scientific analogies; this operational investment leads to the confirmation or refutation of the concepts in use, the hypotheses and theories prepared for each research project, and, ultimately, the general principles set out in the theoretical pole; b) complementary statement: involves subsidiary but essential operations, which even integrate observation, such as description, content analysis, indexing, and authority control; all these operations contribute to the (re)organization of information understood in a strict sense, highlighting the intrinsic properties and intervening in the information process (memory, transfer, and use).

On the ***morphological pole***, the results of the research carried out are formalized through the representation of the object of study and the description of the whole research process which enabled the scientific construction around it. It deals with the organization and presentation of data, objectively checked on the theoretical and the epistemological poles, what shows the interactive character of the quadripolar method of research.

Since the epistemological (paradigmatic reference) and theoretical poles decisively shape the applicational component of IS, which is fully embodied in the technical and morphological poles, it is evident that the ongoing paradigm shift, and the theories adopted (System Theory in this case) require a revision of the technicist framework that has been the essential foundation of the applied disciplines (as archivistics) now integrated into IS. The quadripolar method implies a holistic vision and a dynamic research process in continuous evaluation and refinement, a perspective that gives meaning to the construction of scientific knowledge.

The quadripolar research method is, at our view, the most suitable device for the knowledge of informational phenomena, as it does not limit itself to a merely instrumental view. The foundations of this proposal are explicitly anti-positivist and adjust themselves to the demands of research that is both possible and advisable in the field of human and social sciences. It is also important to emphasize that this proposal recalls the essential idea of 'method' as a complete set of steps necessary for identifying the problem, formulating hypotheses, adopting theories and/or models, verifying these through operations or technical procedures like observation, surveys, and data quantitative analysis (statistical), semi-structured interviews, content analysis, etc., and the final formalization of results. Thus, method refers to a reflective and critical discourse about the research path, not the instrumental and reductive sense used to explain the concrete procedures employed.

Another important advantage of the Quadripolar Method is that it was conceived as a non-linear (a sequence of steps) approach. Unlike Quivy and Campenhoudt (1998) seven stages, the application of quadripolarity can only be in a spiral and there is also an interactivity and simultaneity between the poles, which is a characteristic of exploratory work on human and social phenomena and problems.

Evaluation as a methodological operation: a new theoretical-practical model

The works and studies on the evaluation (or appraisal) of informational flow have been, as mentioned above, primarily developed within the field of archivistics. However, it is our understanding that evaluation can (and should) be extended to all contexts that generate, produce, and accumulate information. That is, evaluation can be carried out within any active or deactivated system, following the same methodological procedures.

A detailed analysis of the issue of appraisal in archives, or how this pragmatic task – driven by the urgent need to eliminate documents to free up space – has been carried out, is beyond the scope of this text. However, it is important to briefly outline the assumptions on which this task has been based so that the proposal we present for a new theoretical-practical model can be better understood.

Archivists' concerns with appraisal were more strongly felt during the period between the two World Wars. These concerns arose due to the significant increase in documentary production during and after the Great War, which led both administrations and archivists to face the problem more decisively than before. Some national trends emerged, with different approaches to this issue. For example, in England, the traditional approach was to destroy documents that had no legal, historical, statistical, economic, or official value. It was understood that archivists should not be involved in the destruction of documents, as it was not part of their responsibilities. The task of elimination was to be left to the administrations themselves (Kolsrud 1992).

On the other hand, the German tradition prioritized preservation over elimination. Consequently, it was believed that administrations should not be the unique judges when it came to document selection, as archivists should have a decisive influence on appraisal. This is a position opposed to the English one, which left decisions to the documents' producers (Kolsrud 1992).

These essentially empirical positions were confronted, starting in the 1950s, with a more systematized perspective that sought to introduce theoretical justification into the guiding principles of appraisal. This position was defended by the American Theodore R. Schellenberg in a study that became the primary reference for American archivists (Schellenberg 1956). The author established a primary value (for the producing entity) and a secondary value (for research purposes). Additionally, he introduced the distinction between "evidential value" and "informational value," as two aspects of secondary value, which to some extent incorporated the criteria used in Germany to guide appraisal. In fact, this American perspective represents a synthesis of the English and German positions and seeks to achieve a very specific objective: to preserve as much information as possible while keeping a minimal number of documents.

Schellenberg's perspective, although subjective since it does not allow for a scientific and rigorous determination of the criteria guiding the attribution of secondary value, has been undeniably important. Since the mid-20th century, it has shaped appraisal in archives. Despite being challenged by various authors, it is still the expression of the dominant paradigm in archivistics and continues to guide appraisal practices worldwide, even being enshrined in manuals and dictionaries considered "classic" in the archival field.

The criticisms of Schellenberg's positions have been taken up in recent years by different "schools" of archival science, with particular emphasis on Canadian schools, some in the United States, and various European "trends." The journal *Archival Science* has given expression to the most recent perspectives in archival theorization and, consequently, has also addressed the issue of appraisal, making it an important reference for understanding the changing perspectives on Schellenberg's ideas (Archival Science 2001-2003). However, despite these attempts at theorization, it seems that, at its core, there is no break with the traditional paradigm, and from a theoretical-epistemological point of view, there are no foundations developed that distance archival science from the technical-custodial phase in which it has been since the end of the 19th century (Ribeiro 2001).

Thus, we begin with the studies developed in the field of archivistics to analyze, expand, and reformulate an aspect of appraisal that, from a theoretical-methodological point of view, can be considered the first and the most essential within the research activity of IS. Studying and monitoring, with theoretical-practical solutions, the creation, circulation, storage, retrieval, and dissemination of information, understood as a human and social phenomenon (and process) for which we already have an operational definition, is the core mission of IS. It is not surprising, therefore, the importance we attach to the evaluation of informational flow, as one of the key operations of the quadripolar method's technical pole.

We will present the specific *modus operandi*, but first, it is important to highlight some basic assumptions without which the application perspective we propose cannot be understood.

Firstly, we do not restrict the flow of information to the administrative and bureaucratic processes emphasized by archival science when it underscores the need for appraisal of accumulated documentary masses and defends the so-called "three ages theory" (current, intermediate, and final or historical) as a method and criterion for selection and sorting. Any type of information, as long as it is created and flows in one or more organizational or combinatory contexts, should be evaluated using the same framework (with additional indicators and categories). Whether dealing with documentation produced, received, and accumulated by a department, division, or office of any governmental entity (central or local), we can and should apply the same evaluation framework as that used for documentation created and circulated electronically or in print on a limited range of specialized topics relevant to a manufacturing unit, laboratory, or commercial entity. And by applying the proposed procedures, we will reach a decision on what should remain in the corresponding memory.

Secondly, the classic and restrictive idea that we can only eliminate internal and official documentation produced by administrations (such as correspondence, accounts, reports, etc.) and not the documentation that is edited, collected, and accessed in libraries and documentation centres makes no sense in the context of the paradigm imposed by or shaped by the Information Society in which we live and are increasingly immersed. This paradigm requires a radical rethinking of the heritage conception (static and supposedly autonomous from the daily activities of social and historical actors) that archival science continues to claim as its legacy and standard-bearer through the "three ages theory." Documentation produced and accumulated must be appraised – and this point is consensual – but not as an end in itself determined by almost exclusively historicist interests. A complete shift in this perspective is necessary, and this implies embracing evaluation as a key element of a broader and more consistent theoretical-methodological framework. It also implies recognizing that the evaluable object is not something merely tangible and physically scattered across miles of shelves, cabinets, and storage rooms but the natural and endless product of the human mind (phenomenon and process) referred to by the term/concept "information".

Thirdly, the dynamic criteria that give meaning to informational flow are centred on the memory of any entity that produces, receives and accumulates information. For this reason, the concept of institutional or organizational memory (as vital as personal or individual memory) takes on enormous and fundamentally prospective importance. Memory is intimately linked to information, focusing on the present and future: preserving and accumulating information makes sense only to generate, stimulate, and qualify action today and tomorrow. All memory is selective, and organizational and institutional memory must be as well. But selection happens here through two ways: either by randomness, determined by the pressures of everyday life, or by a sense of belonging or pertinence dictated by the essential, structuring, and long-term objectives of the organization or entity. The first route is quite common and characterized by immediacy, lack of planning, and improvisational and opportunistic management. The second one is increasingly recommended by management gurus under various names, expressions, and models, and it truly represents the only guarantee of achieving innovation sustained by experience and accumulated information, filtered through the guiding lines of the long-term mission.

Fourthly, we do not find it consistent or feasible to evaluate informational flow without "configuring" or "focusing" it through a theory and models designed and developed to achieve good and better results. Our preference for system theory is subject to the ongoing demonstration process in different study, organizational implementation, and research situations. The balance so far, though still in its early stages, is encouraging. Therefore, we find it very stimulating to draw conclusions – necessarily provisional and reversible – on the evaluation of flows within various systemic frameworks – flows in different organizations and entities (individuals, associations of all types, small businesses, etc.) and in non-organizational conditions (so-called combinatory systems subject to the tension and interaction of micro and macro-behavior).

Outlined these basic assumptions, we then proceed to present the evaluation model we developed, which we have already had the opportunity to use and test in various situations along the last two decades, with the publication of the results in some cases (Carvalho 2002; Ribeiro and Fernandes 2003; Fernandes 2004). The fundamental principles of the emerging paradigm of IS, previously summarized, necessarily have several implications for the development of studies and the technical application work, and naturally influence the procedures inherent to appraisal. As we have seen, evaluation/appraisal is one of the methodological operations situated within the technical pole of the quadripolar method.

In fact, applying the quadripolar research method, emphasizing its main operations as previously outlined, focuses on retrospective and prospective analysis/evaluation. Here, *organic-functional analysis* is crucial, as it is an indispensable requirement for achieving a precise understanding of the system's structure and the functions/competencies of its various components. This is the way to make it possible to accurately characterize the production context of the informational reality under study. Often, this analysis must be carried out diachronically to correctly understand the changes in the structure and functions that have shaped the information over time.

Furthermore, analyzing the functional component of the system leads to the implementation of certain operations as regular "prophylactic" measures aimed at optimizing the functioning of the information system itself. Evaluation, a fundamental operation that allows the detection of redundancies and informational "waste," falls into this category and is the focus of our discussion here.

Based on these assumptions, the proposal of a new model to frame the procedures inherent to the evaluation of information flow rests on several criteria and parameters (see Appendix), which we outline as follows:

Pertinence: Literally meaning belonging to the action of someone or some entity, pertinence can be measured, in informational terms, through the triad of essential objectives – reason to exist + organizational structure and competencies/functions + memory – graded at three levels (A, B, and C) corresponding to a direct, indirect, or peripheral relationship between informational acts and the aforementioned triad.

Density: Literally the quality of being dense, thick, or compact, density implies, in informational terms, determining whether an act or document is primary/original, with or without duplication/exact copies, or secondary (a summary or synthesis, a part or accumulation of primary/original information), also with or without duplication/copies.

Frequency: Defined as the repeated occurrence of acts, frequency is understood here as the quantification of the periodicity of information use/access, whether during the production/reception phase (the genesis or decision-making phase, also called the current or administrative phase) or in the immediately subsequent phase (the stable, post-genesis, and post-decision phase, which is perennial and definitive, as well as progressively more open to external access to the information system). The results obtained in both phases can provide a comprehensive understanding of whether an "intermediate use" exists (a highly debatable notion) and whether it is true or not that the administration completely loses the need for access to information with more than forty years old.

The first parameter – *pertinence* – requires an exhaustive and in-depth collection of data essential for a structural (organic-functional) characterization of the organization (or entity), which is often illustrated through organizational charts and flowcharts. It also demands a thorough analysis/evaluation, which the IS researcher or IS specialist must undertake. This role positions them as the most suitable interlocutor in a broad interdisciplinary framework, where, for instance, administrators, managers, and other stakeholders in the informational and organizational process play a crucial role. Constructive and dynamic dialogue with these and other agents should be a regular and standardized practice, conducted both informally and formally: easy and spontaneous availability for interviews, responding to surveys, and participating in joint research; and, more formally, integrating working groups.

Within this parameter, three levels are considered:

- *Level A* includes all information directly related to the essential and usually immutable objectives of the entity responsible for the information system. These objectives are typically embedded in its foundational "constitution" and fulfilled through an organic-functional structure that evolves over time but remains consistently focused on achieving these original goals in daily operations.
- *Level B* encompasses information related to what are commonly referred to as "support functions," such as information produced/received in the management processes of the organization to meet the objectives that justify and legitimize its existence. This includes accounting and financial aspects, human and technical resources management (equipment, support materials, etc.), and a variety of external contacts essential to regular and goal-oriented activities. This level is present in any organized or operational information system.
- *Level C* includes certain informational acts that are objectively marginal or peripheral to the objectives or to the "support operations" of management. It also includes the redundant quantity of documents and series resulting from outdated, routine, and irrational administrative and executive practices still common in bureaucratic, hierarchical, and stagnant organizational models. Inclusion in this third level often indicates the concrete possibility of elimination, although this may be hindered by exceptional results from the *frequency* parameter.

The second parameter – *density* – also plays a decisive role in the selection of institutional or organizational memory. A primary/original informational act renders its exact duplicate unnecessary, while a secondary act (summary or synthesis, part, or accumulation of primary/original acts) can often be replaced by the primary, except in the case of summaries/files/databases or cumulative types, such as annual reports, which aggregate dispersed information from monthly, quarterly, or semi-annual reports.

Another category of information is *progressive*, which includes certain types of information related to specific functions (scientific and commercial) subject to periodic changes or updates. For example, a commercial prospectus is valid for a campaign lasting months or one to two years, after which it becomes outdated, which makes its elimination inevitable. However, its relation to the pertinence parameter should also be considered to estimate the period of temporary conservation. Therefore, it is both convenient and mandatory in this and all cases to "cross-reference" or relate parameters and levels.

Finally, the parameter *frequency* (rate) of use is considered. This is based on a standard average of 52 requests per year, equivalent to one per week. Below this average, usage is classified as weak or minimal (if fewer than 20 requests), while above it, usage is categorized as medium or high. The interpretation of the rate varies depending on factors such as user identification, justification for the request, and the occurrence of requests within a concentrated or short period (six months, three consecutive months, one month, two weeks, or one week). The results and their implications do not directly influence the choice of eliminable information, as low usage may correspond to a set of primary documents (series or otherwise) of Level A, which require permanent preservation. However, these results are indispensable for understanding the informational process in its intrinsic properties and how organizations "manage" their own memory.

The indicators or parameters outlined must be emphasized and cross-referenced to achieve a global and scientific "reading," analogous to the statistical insights already obtained through infometric or bibliometric laws. By assigning a weighting factor (1 or 0, where 1 indicates information to be preserved and 0 indicates information that can be discarded) to the three parameters that guide evaluation, it is possible to establish a matrix that typifies the intersections between these parameters. This enables objective decision-making regarding the retention or elimination of informational memory (see Appendix).

The operationalization of this model has already been tested in various contexts. For this purpose, data collection forms and analysis grids were developed. The results of this practical application have been published in some articles, aiding in the clearer understanding of specific operational procedures. Since it is beyond the scope of this text to develop the practical component of appraisal, readers are referred to available works on the subject (Carvalho 2002; Ribeiro and Fernandes 2003; Fernandes 2004).

Conclusion

To conclude, it is essential to emphasize that evaluation is not exclusive to IS or other social sciences. Nor is evaluation merely a specific task or performance carried out as if following a predefined recipe. Evaluate is, first and foremost, a verb that acquires different meanings depending on its context of application. In a scientific-technical framework, evaluation cannot stand alone as an act or, more accurately, as an operation. This core idea underpins the entire development presented in this discussion: to frame evaluation as one of the pivotal operations within the technical pole of the quadripolar method, tailored to qualitative research in the social sciences (and, consequently, in IS), and to integrate application areas previously treated as independent and fragmented.

Evaluating the performance of services (many institutionalized, others increasingly provided through IT and telematics), the quality (effectiveness, efficiency, and speed) of information retrieval tools (satisfaction or failure in controlled searches, especially in search engines on the Internet), and the circulation, selective retention, and recovery of informational flow – ubiquitous across all sectors, groups, and institutions of human society worldwide – are not distinct activities. They are integral components of a single technical operation, preceded by observation and, potentially, experimentation, culminating logically in analysis.

Among other advantages, the quadripolar method enables technical operations and procedures to be situated within interactive, interconnected polar cores, fostering a productive and effective spiral of scientific advancement.

Through the *epistemological pole*, any individual who studies, investigates, and tests theoretical-practical solutions in the field of IS is compelled to reflect on and understand their limitations. This pole reveals the subtle yet fundamental paradigm shift – essentially, a change in perspective or in the psychological, formative, economic, and socio-political framework.

The *theoretical pole* also precedes and guides the technical procedures of the method. Given that elements obtained through observation, results from various experiments, and evaluation/analysis findings contribute directly to (re)formulating hypotheses and consolidating theories, it is clear that without hypotheses and theories, the scientific study and resolution of any properly identified problem become obstructed or significantly distorted.

Without the *technical pole*, it is impossible to finalize results and ensure their dissemination, discussion, revision, and impact on future research or new research projects. These are facilitated through the *morphological pole*, but it must be clear that evaluation/appraisal as an operation does not occur in IS in an isolated manner or without the intent of achieving a comprehensive explanation or a well-founded practical resolution at the theoretical level.

At a more operational level, the researcher must adopt the role of a constant and continuous creator of tools, fully justified in one project but potentially irrelevant in another. This diversity of technical tools – understandable and desirable – can take on multiple "formats and tones," yet it fits within the operations of the technical pole, focusing on the phenomenon and process of information in its entirety.

It is, therefore, necessary to distinguish, in the context of evaluation, the feasibility of a uniform matrix designed and applied in the field of informational flow (see Appendix). There exists, consequently, a unity of the object of study and, simultaneously, without any paradox, a diversity of resources within the methodological framework used to study it.

References

Archival Science: international journal on recorded information. 1, 1 (2001) - 3, 2 (2003).

Bertalanffy, Ludwig von. 1976. *General System Theory: Foundations, Development, Applications*, revised edition. New York: George Braziller.

Carvalho, Cristina. 2002. "A Avaliação em Arquivística: um estudo de caso em ambiente electrónico". *Páginas a&b: arquivos e bibliotecas*. 9 (2002): 7-38.

De Bruyne, Paul, Jacques Herman, and Marc de Schoutheete. 1974. *Dynamique de la recherche en sciences sociales de pôles de la pratique méthodologique*. Paris: P.U.F.

Fernandes, Daniela Teixeira. 2004. *Pedra a pedra: estudo sistémico de um arquivo empresarial*. Lisboa: Gabinete de Estudos a&b.

Kolsrud, Ole. 1992. "The Evolution of basic appraisal principles: some comparative observations". *The American Archivist*. 55, 1 (Winter 1992): 26-39.

Lessard-Hébert, Michelle, Gabriel Goyette, and Gérard Boutin. 1994. *Investigação qualitativa: fundamentos e práticas*. Lisboa: Instituto Piaget.

Mella, Piero. 1997. *Dai Sistemi al pensiero sistemico: per capire i sistemi e pensare con i sistemi*. Milano: Franco Angeli.

Quivy, Raymond and Luc van Campenhoudt. 1998. *Manual de investigação em ciências sociais*. 2ª ed. Lisboa: Gradiva.

Ribeiro, Fernanda. 2001. "Archival Science and changes in the paradigm," *Archival Science: international journal on recorded information* 1, 3 (2001): 295-310.

Ribeiro, Fernanda and Maria Eugénia Matos Fernandes. 2003. "O Sistema de informação arquivística da Universidade do Porto: potenciar o uso da memória informacional retro-prospectivamente," in *Archivos universitarios e historia de las universidades*, edited by José Ramón Cruz Mundet, 279-299. Madrid: Instituto Antonio de Nebrija de Estudios sobre la Universidad, Universidad Carlos III de Madrid; Editorial Dykinson.

Ribeiro, Fernanda and Armando Malheiro da Silva. 2004. "A Avaliação de informação: uma operação metodológica," *Páginas a&b : arquivos e bibliotecas*. 14 (2004): 7-37.

Schellenberg, T. R. (1956). *Modern archives: principles and techniques*. 2nd ed. Chicago: University of Chicago Press.

Silva, Armando Malheiro da and Fernanda Ribeiro. (2014). "Information evaluation: a scientific and comprehensive approach," in Fernanda Ribeiro, Armando Malheiro da Silva, and George Jamil, *Rethinking the conceptual base for new practical applications in information value and quality*. 19-41. Hershey: IGI Global, 2014.

Silva, Armando Malheiro da and Fernanda Ribeiro. 2002. *Das "ciências" documentais à Ciência da Informação: ensaio epistemológico para um novo modelo curricular.* Porto: Edições Afrontamento.

APPENDIX

Evaluation parameters and levels applied to the information flow

Identification of the produced/received/accumulated information and its contextualization:

OBJECTIVES	ORGANIC-FUNCTIONAL STRUCTURE	COMPETENCES ATRIBUTIONS	ACTIVITIES (procedures / tasks)	SERIES and/or INFORMATIONAL TYPES
(...)	(...)	(...)	(...)	(...)

Evaluation parameters:

PERTINENCE (to belong, in informational terms, to somebody's or some entity's action)
Level A (information directly related to the objectives / organic-functional structure / memory)
Level B (information indirectly related to the objectives / organic-functional structure / memory)
Level C (peripheral and/or not up-to-date information considering the objectives / organic-functional structure / memory)

Weighting: **1** (level A or B information) or **0** (level C information)

DENSITY
Primary information
Secondary Information (produced from primary one)
- Partial
- Summarized
- Cumulative

Progressive information (in S&T)
Duplicated information

Weighting: **1** (more dense/complete information) or **0** (less dense/complete information)

FREQUENCY (of use)
Maximum/medium use - once or more than once a week
Minimum use - less than once a week

Weighting: **1** (maximum/medium use) or **0** (minimum use)

Analysis of series and/or informational types

PERTINENCE:
Level A ___ **Level B** ___ **Level C** ___
Weighting ___ (**1** or **0**)

DENSITY:
Establishment of informational relations to determine <u>density</u>:

INFORMATIONAL RELATIONS

Information to be analyzed	**Related information** (1)
Primary information __ Duplicated __	→ Primary information ________________ Primary and duplicated information ________ Secondary information ______ Partial ________________________ Summarized _______________________ Cumulative ________________________ Progressive information (S and T) ________
Secondary information __ Partial __ Summarized __ Cumulative __ Duplicated __	→ Primary information _________________ Secondary information _______________ Secondary and duplicated information_____ Progressive information (S and T)_________
Progressive information (S and T) Duplicated __	→ Primary information _________________ Secondary information ________________ Progressive and duplicated information _________

(1) Mention, in each case, the reference of series or informational type related to the information to be analyzed, referred to in the first column

Weighting ___ (**1** or **0**)

FREQUENCY:

Maximum/medium use – once or more than once a week - **1**
Minimum use – less than once a week - **0**
Weighting ___ (**1** or **0**)

- Decision about the final destination of the information:

PERTINENCE	DENSITY	FREQUENCY	FINAL DESTINATION
Level A	1	1 or 0	Permanent conservation
Level A	0	1	Temporary conservation
Level A	0	0	Elimination
Level B	1	1 or 0	Permanent conservation
Level B	0	1	Temporary conservation
Level B	0	0	Elimination
Level A	1 or 0	1	Temporary conservation
Level C	1 or 0	0	Elimination

Note: deadlines must be established for temporary conservation of information (schedule of selection) according to frequency of use

9

The Robotics Project: Collecting Complex Science at Carnegie Mellon

Julia Corrin and Kathleen Donahoe, Carnegie Mellon University, United States

Abstract

In 2019, Carnegie Mellon University (CMU) launched The Robotics Project, an interdisciplinary collaboration between the University Libraries and the School of Computer Science to develop best practices for archiving robotics materials. Since the project's inception, archivists at Carnegie Mellon have worked to establish guidelines for collecting and processing complex scientific collections that diverge from traditional archival materials. The multimodal nature of these records necessitates subject and technical expertise beyond that of a typical archivist, requiring additional research and early engagement with potential donors. To address these challenges, the Carnegie Mellon University Archives has published a white paper and established the Digital Robotics Archive to serve as a prototype digital repository for selected materials. As robotics technology advances, the project aims to preserve historical and contemporary research contributions for future generations.

History of Robotics at Carnegie Mellon University

Carnegie Mellon University was founded in 1900 by industrialist Andrew Carnegie as Carnegie Technical Schools, offering advanced professional training in industrial fields for local Pittsburgh residents. In 1912, it became the Carnegie Institute of Technology, later merging with the Mellon Institute in 1967 to form Carnegie Mellon University.

CMU emerged as an early leader in computer science and artificial intelligence, with foundational contributions from Alan Perlis, Herbert Simon, and Allen Newell. In 1956, Simon and Newell introduced the first working AI program, the Logic Theorist, at the Dartmouth Conference on Artificial Intelligence, a seminal event that marked the beginning of AI as a scientific field.[1] This breakthrough laid the groundwork for subsequent AI developments, influencing research directions at CMU and beyond. The establishment of the Computer Science Department in 1965 signified an institutional commitment to this work, eventually leading to its recognition as a standalone school in 1988. By 2018, CMU further solidified its investment in the field by introducing the first undergraduate degree program in artificial intelligence.

In 1979, faculty from multiple disciplines, including Computer Science, Electrical Engineering, Mechanical Engineering, Mathematics, Civil Engineering, Psychology, and the Graduate School of Industrial Administration (now the Tepper School of Business), collaborated to establish the Robotics Institute—the first academic robotics program in the United States. Research areas have encompassed locomotion, autonomy, manufacturing, computer vision, and manipulators, each contributing to significant advancements in robotics applications. The Robotics Institute lists four key research themes that, when integrated, typify modern robotics: sensing and perception; thinking about actions; interacting with the world; and robot embodiment.[2]

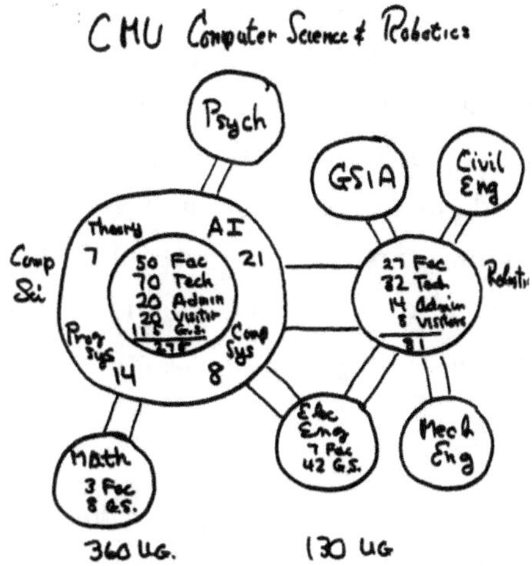

Figure 1[3]

A diagram by Allen Newell showing the breakdown of faculty and graduate students in Computer Science and Robotics at Carnegie Mellon, and their research backgrounds.

The Field Robotics Center (FRC), founded in 1983, pioneered research in autonomous robots for hazardous environments, such as disaster sites and deep-sea exploration. Notable projects included developing self-driving Navlab vehicles and Ambler, a walking robot designed for planetary exploration. Robotics research at CMU has relied on a combination of federal and corporate funding, often complicating intellectual property rights. In 1996, CMU established the National Robotics Engineering Center (NREC) to commercialize robotics research, partnering with companies such as Caterpillar and Uber and managing classified projects for NASA and the Department of Defense.

Call to Action

Because of its institutional significance, the Robotics Institute should have been a key curatorial area for the University Archives. However, during a survey of the Carnegie Mellon University Archives holdings in 2018, archivists at CMU discovered that there were only 1.5 linear feet of "official" robotics records, the majority of which were professional publications from a former director of the institute.[4] If this department's records existed, they were spread amongst other collections, and no notations of these related materials existed.

With the first generation of academic roboticists starting to retire, it was clear that the Archives needed to act quickly to fill these gaps. However, it became apparent very soon after that traditional approaches to archival collecting would not work in this case. During our initial conversations with a small group of senior roboticists, we used a conventional archival framework, focusing on the future research value of their papers to convince them to donate or transfer material. These early conversations were, for the most part, unsuccessful. We identified issues with our early approach that helped reshape our future conversations.

Archival Stereotypes

Some roboticists did not know what an archive is, and those who did subscribed to the "dusty box" stereotype. They envisioned any donation to the archive lingering unused and unseen on shelves. They also failed to recognize the difference between archival collecting and museum collecting.

For these scientists, the physical object was the record of their work. And the object needed to be seen and touched to be understood.

Parts Cannibalization

We started by approaching the most senior researchers and those who had been at Carnegie Mellon the longest. While almost all of them were concerned about the legacy and importance of their early research, much of that work was already gone. Due to the expense of robotics parts, early projects were often cannibalized and incorporated into new work, so even if we were willing to expand our understanding of what an archive collects, a large part of the physical history no longer exists.

In some cases, because the physical robot was considered the final research output, researchers had not always maintained the project files in which we were interested. In several instances, research files had been lost, actively destroyed, or damaged by poor storage. When records had been kept, it was challenging to communicate to roboticists why someone might be interested in them.

Team Science

We framed our initial list of prospects with the archival principle of provenance in mind. In our conception, the lead researcher on a project would hold the central project files. Based on that assumption, if we could work with individuals who lead major labs, we would be able to capture records on a broad range of robots, and do so quickly.

But robotics is a team science.[5] Research projects often rely not only on a large team in one lab, but also on the collaboration of multiple labs. Each robot may, therefore, represent the culmination of multiple related research agendas. For example, understanding Terragator–an autonomous 6-wheeled robot capable of leaving the laboratory setting–requires documenting not only the work of the Field Robotics Center, but also each lab or group that used the robot for their research.

Documentation Standards

We also quickly learned that the documentation standard for robotics was video of the robots operating. While in many ways this was good news, as the management of AV items is already part of standard archival practice, it also promised challenges. The Robotics Institute already managed a collection of thousands of videos that allowed for rudimentary search and access. However, the preservation masters held approximately ¼ petabyte of data, far beyond the Archives' storage capacity.

AV assets also present barriers to accessibility and can require significant resources to process fully. These videos are especially complex, as most contain no audio to aid description.

The Robotics Project

The mistakes in our initial approach to documenting robotics forced us to step back and revisit our goals. It was clear that a traditional archival approach alone would not be practical—it would not provide us with the buy-in we needed, leave key aspects of the work absent, and ignore the discipline's inherently physical nature.

We first sought to understand how others approached robotics as a collecting field, but we found little outside museums and nothing that looked at research holistically. We also investigated other disciplines that are particularly difficult to document, including dance. From this cursory research, members of the Carnegie Mellon University Libraries, including Katherine Barbera and Brian Matthews, decided to assemble an interdisciplinary team that included archivists, digital humanists, historians, and anthropologists to understand the field of robotics better and identify potential new approaches.

With funding from The Alfred P. Sloan Foundation, the University Archives hired a processing archivist to investigate how one archives a robot. The grant had two core goals: a white paper providing advice on collecting and processing complex scientific collections, and a sample digital collections site for some of the materials acquired during the pilot phase.

Ethnographic Research

Team members started their research by conducting an ethnographic survey of the robotics labs at CMU to understand the research enterprise better. While multiple units on campus, including Engineering, conduct robotics research, the project team decided to limit the initial scope to the Robotics Institute. The Robotics Institute has approximately 42 labs and centers, and the breadth of work being done within and across these research groups crosses multiple subfields: field robotics, autonomous machines, computer vision, microdynamics, biorobotics, and much more. To understand the needs of these groups and the potential outputs that may be collected in the future, the project team created a sample group of labs with different disciplinary focuses to give us as varied a picture of potential archival outputs as possible. In addition to interviews with individual members of the labs, there were also scheduled observation sessions where possible. This endeavor aimed to discover products and by-products of the research process, the contexts of where and how these products were created, and if there were materials considered at-risk for long-term preservation. The information collected from these surveys was intended to inform the development of a CMU-specific collecting policy for these types of records.

The responses to the survey further emphasized that the project team was too focused on archival topics and jargon, which often failed to translate to non-archival professionals. For example, the phrase "archive" elicited thoughts of museums and visitor centers with robots on active display rather than a site for research or an information corpus. The word "documentation" was also confusing, as each discipline used the term differently. These instances, among others, forced a review and revision of how we phrased questions during later interviews.

After the interviews and surveys were complete, the transcripts were used to extract data for the final toolkit. We categorized statements from participants into different areas of interest for our research, such as Products, Obsolescence, Culture, Connections, and Ecosystem. Once other members of our team completed this encoding work individually, the results were compared, and the commonalities were used to help shape the final output.[6] For example, several participants mentioned professional literature as an essential part of their research process and one of the project's end goals. We categorized these literature mentions and materials, like conference proceedings, data sets, or physical robots, as Products. This fell in line with expected results, since preliminary conversations with roboticists at the beginning of The Robotics Project had indicated that these materials were often outputs of the research cycle (see Fig. 2). The types of materials that were listed as Products helped determine what we would emphasize in our collecting work with donors.

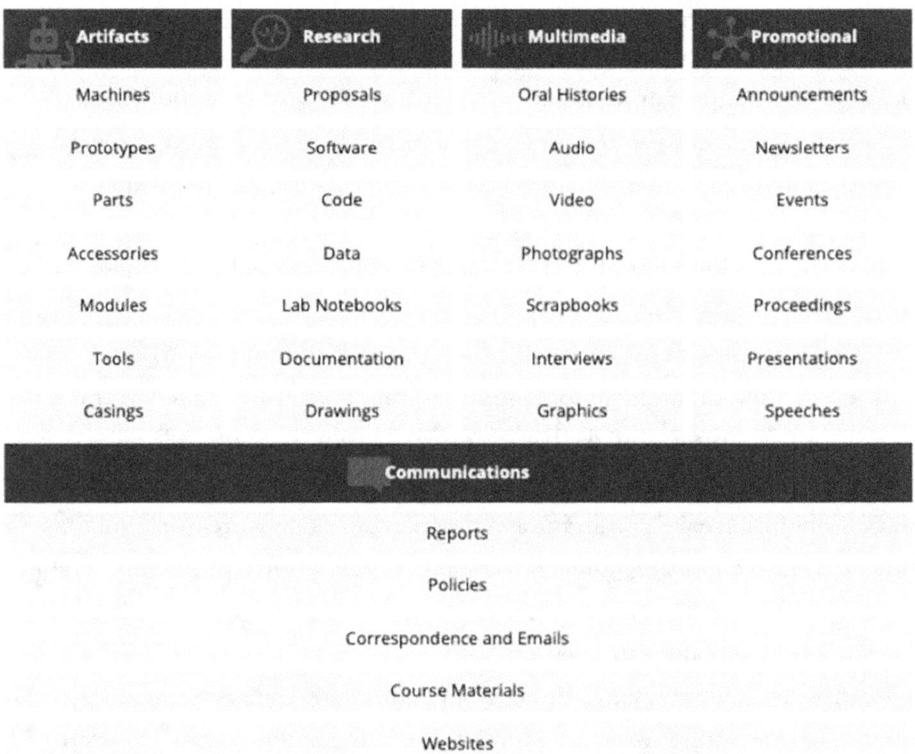

Figure 2. The different categories of outputs discovered through initial conversations with roboticists at CMU. [7]

The labs also reported using different platforms for group work and research, indicating a lack of standardization across the Robotics Institute. From an archival perspective, this means that of the approximately 42 labs in operation, there might be at least 42 different repositories, tools, or storage services where documentation and records of a robotics research project might exist.

In addition to informing collection development decisions, the findings from the ethnographic survey and meetings with consultants on digital preservation, object conservation, and inclusivity in the archives were used to create a conversation guide for building relationships and collections through the pre-custodial model of archiving. The experience of the archivists on this project indicates that for archival collections in complex, multimodal fields, some interactions with donors may require more intense research and preparation. The ROBOTS (**R**esearch, **O**utreach, **B**uild, **O**ffer, **T**ransfer, **S**teward) conversation guide is meant to provide steps to help archivists complete this early preparation, and includes descriptions and sample questions to consider for each of the six steps.[8] The steps listed should not be considered a requirement; they are, more specifically, an example of what worked for The Robotics Project throughout the survey and what might work for archivists facing similarly complex collections at other institutions.

Digital Collections

Pre-grant conversations with roboticists also revealed a strong interest in making materials accessible online (and not, as per stereotype, in dusty boxes), and high expectations for the final capabilities of such a product. As the University Libraries were already migrating existing digital collections into Islandora 8 in the fall of 2021, it was decided to develop a prototype robotics portal on the same platform. With Islandora 8, the Robotics Project could use new linked data features. Linked data was a critical feature because one of the core goals of the site was to demonstrate the interconnected nature of robotics projects and the field as a whole.

While research grants typically have a principal investigator (PI) who leads the project, the project team learned that it is often difficult to point to one person as a "creator" of robotics research. A PI for a project may have a handful of staff working with them, a cohort of graduate students studying with them, and, on some occasions, undergraduate students working in the lab. On larger projects, multiple labs might work on different aspects of the project to create different end products. Where initially an archivist might assume five to ten people were affiliated with a project, in that latter example, there could be 30-40, and each could potentially have records or materials related to their work. The digital collections site needed to demonstrate these relationships between researchers and research products to represent this web of connections best.

For the prototype, items were selected from the initial collections acquired as part of The Robotics Project. These included the Manuela Veloso Papers, the Christopher G. Atkeson Collection, the Jodi Forlizzi Papers, the Robotics Institute Records, and the Robotics Institute Photograph and Video Collections. Materials were selected based on novel formats and those that document the same research project differently.

The Veloso, Atkeson, and Forlizzi collections all contributed robotic artifacts, including multiple versions of "soccer robots" designed to test robotic cooperation, Pearl - a healthcare aid that was part of research into robotic assistance for elderly people who may be suffering from chronic disease, and Snackbot - a mobile delivery robot that used to ferry snacks between researchers' offices and helped with the study of human-robot interaction. The Veloso Papers also provided some sample documentation in the form of game reports from the RoboCup World Cup event, where soccer robots were tested against teams worldwide. The Photograph and Video Collections include materials created by and for the Robotics Institute: photographs and videos documenting different moments in the research process for some projects. The Robotics Institute Records include the research reviews produced over 12 years by scientists from within the institute, documenting the different fields where work occurred. Approximately 65 objects were selected from these collections to represent what might be possible with robotics collections. The site includes photographs, video footage, oral history interviews, digital surrogates of physical robotic artifacts, and PDFs of relevant documents.

Many digital collections sites provide access centered around collections. Because of the complex relationships discussed throughout this paper, the team determined that landing pages for people, projects, and topics would be a priority. A landing page for an individual includes a description, a list of the projects or teams they may have been part of, a list of colleagues based on those project/team lists, and links to relevant objects and documents. The project pages include a brief description, a list of team members, relevant subjects or themes like "autonomous robots" or "artificial intelligence", and links to related objects and documents on the site. Each topic or theme also has a landing page with links to relevant items. For the prototype, many of these entry points required developer intervention to create; however, a larger corpus of data might make populating this information easier in the future.

The networks and connections on the site are not all-encompassing and reflect the relationships in the digital objects, not the full scope of the actual research networks and projects. Some people worked on projects and colleagues of named individuals who are not listed because none of the ingested objects represent them. The site may have a more in-depth description in the future, and with additional sources ingested. For the Sloan grant, however, attempting to recreate the entire social network of the Robotics Institute and Carnegie Mellon robotics in general was beyond the project's scope.

Outcomes

Since the end of the Sloan Foundation grant in 2022 and the publication of the white paper in 2023, The Robotics Project has continued to grow. The University Archives has added multiple collections to its holdings, of varying sizes, related to robotics research at CMU. Outreach efforts also continued, with the Archives contributing significantly to an anniversary celebration of the Field Robotics Center (FRC) in 2023. The two-day event brought alums and current and former staff together to reminisce and discuss current topics in field robotics. During the event, archives staff presented materials from the archive for review. Also, it used the opportunity to ask attendees to assist with the identification of people and projects in photographs. This was also an opportunity to get first-person accounts of FRC from different eras; short, 5-10 minute oral histories allowed participants to talk about their favorite projects, what brought them to CMU, and more. These videos then became part of the archive, with the aim of sharing them online in the future.

In addition to the FRC event, the Archives is working with Robotics Institute, School of Computer Science, and College of Engineering personnel to develop and implement a permanent, long-term display in a soon-to-be-finished building off-campus, the Robotics Innovation Center (RIC) in the Hazelwood neighborhood of Pittsburgh. The work involves several different groups of stakeholders and has allowed the Archives staff to discover more objects and history to share with the public.

Collecting robotics research - and robots themselves - has helped us identify additional challenges and areas for future research. Many of these were predictable (and predicted) but arose much more quickly than expected. Others reflect our naivety about this field of research when we commenced with the work. All offer lessons and cautions for others looking to undertake similar work.

Object Management

Archives are not museums, but at institutions without an object-collecting arm, archives often have to step in to fill the gap. The storage, display, and conservation of robots is significantly more complicated than the modest ephemera that archivists are usually tasked with managing. Most potential accessions for The Robotics Project are complete robots or parts of robots, which are typically not small enough to fit into an oversized box, let alone a record center carton-sized one. Carnegie Mellon's robotics research is also known for its autonomous vehicle research, meaning that some potential donations are cars, trucks, buses, and tanks.

Donations need to be carefully considered due to space restrictions and the ability to move the robot. Some of these objects weigh hundreds, if not more than 1,000 pounds, which requires more logistical strategy than the average archival move. As we move forward, these experiences will help shape and finalize the collecting policy for the project so that we can better manage expectations in the future.

Security & Ownership

Robotics research is expensive, but we assume that, as with other federally funded research, we can collect and eventually provide access to it. Collecting becomes much more complex when security clearances and corporate funding are involved. Much of CMU's robotics research is public and documentable, but the university also receives heavy funding from the Department of Defense (DOD) and corporations, including Caterpillar.

Projects with DOD funding, even older projects, require complex clearances and releases to make any aspect public. Even on visits to warehouses storing inactive projects, team members were told to refrain from taking pictures. Projects done with corporate funding involve complex, proprietary agreements. Rather than risk accidentally releasing confidential information, we are first focusing on documenting projects with more straightforward sponsorship requirements.

Advocacy and Outreach

While the team was always aware that outreach would be essential to launching The Robotics Project, we underestimated the need for ongoing outreach. Robotics faculty and staff needed to be reminded that this is an ongoing project, not a one-time effort. Because this project is a low priority for them, some needed to be reminded of its existence. Faculty especially require regular contact and evidence of concrete project outcomes.

Conclusion and Future Directions

The first phase of The Robotics Project was necessary to build initial connections with robotics researchers and, most importantly, to understand their research process. At the end of the research phase, we knew what *not* to do, but hadn't entirely determined what *to* do. We'd received a few collections and multiple robots and robot components, but the larger concept of the project did not seem to have translated across disciplines. There were still issues communicating value, explaining collecting goals, and containing the scope.

We have, however, had the opportunity to participate in other strategic projects and partnerships that are leading us to revise our approach moving forward. As part of a more significant effort to make actual robots visible on campus, the Robotics Project team was invited to curate and produce a display in the new Robotics Innovation Center, set to open in the fall of 2025. We were challenged to put physical objects in the space, without any dedicated floor space. The display also needed to incorporate both contemporary and historical projects. This led to a mismatch between what the University Archives held in its collection and what made sense for the space.

While earlier conversations with researchers had demonstrated some reluctance to participate in The Robotics Project, we were surprised to find enthusiastic acceptance for this new proposal. Almost every researcher or team we spoke with was more than happy to have their work included and readily offered documentation to contextualize the work. While further study is needed, we've identified three concepts that will enhance our collecting efforts moving forward.

Concrete Outcomes

Roboticists are builders. They create complex physical objects that must complete a specific task or purpose. They do not deal in the abstract. While convincing them of a vague, future use for their papers proved difficult, showing them an immediate use was easy. Tying collection conversations to concrete, visible outcomes is necessary. There needs to be value for the now, in addition to value for the future.

Contemporary Collecting

Focusing on contemporary collecting, rather than retrospective collecting, is also advantageous. In the abstract, researchers are more excited about promoting their current work than past projects. Their research focus may have changed, or they may find their old work outdated. In some cases, they don't feel anyone else would be interested if they are no longer interested in a robot.

And, in a realistic sense, recent projects are more likely to still exist. Robotics research falls victim to both active cannibalization and benign neglect. During our conversations, we've discovered robots disassembled to reuse expensive parts, robots taken apart to use a key component in a new version, robots lost to floods, and robots that are just lost. The documentation – schematics, codes, research reports – often meets a similar fate: discarded to save space, damaged by water, or misfiled and forgotten.

It is important to begin gathering material while the project is still active to ensure the physical object and the contextual material exist.

Revisiting Provenance

We often encountered a complex web of ownership in seeking permission to display specific robots. Even lead researchers were sometimes unsure who would need to approve the donation. This further reinforced one of our initial findings—that robotics is a team science, with multiple projects and research questions operating together.

Early collecting efforts focused on traditional archival structures and theories related to provenance. We targeted collections of individuals or labs, which may have been too large and too fungible a scope. Moving forward, we intend to focus on collecting at the project level, where ownership may be clearer. This will hopefully address both the complex research environment and the need for earlier intervention.

The first phase of the robotics project was speculative. The project team needed to understand how robotics functions as a research field before determining how to approach collecting. The initial collections received during the first phase–the Atkeson Collection, the Veloso Papers, and the Robotics Institute Photograph Collection–helped us understand the documentation landscape and how seasoned researchers think about their work. Moving forward, we continue to try to put these findings into action and begin to confront the challenge of preserving objects that rely on physical and digital interaction to function.

Endnotes

1 Pamela McCorduck, *Machines Who Think: A Personal Inquiry into the History and Prospects of Artificial Intelligence* (Natick, MA: A.K. Peters, Ltd., 2004), 170.

2 "Research Overview," Robotics Institute, Carnegie Mellon University. Accessed December 16, 2024. https://www.ri.cmu.edu/research-overview/.

3 Allen Newell Papers, Carnegie Mellon University Archives. http://doi.library.cmu.edu/10.1184/pmc/newell/box00115/fld08916/bdl0003/doc0001.

4 Our collection survey utilized methodology outlined in *Taking Stock and Making Hay: Archival Collections Assessment*, published by OCLC Research in July 2011.

5 L. Michelle Bennett, et al. *Collaboration Team Science: Field Guide* [Rockville, Md.]: U.S. Department of Health & Human Services, National Institutes of Health, National Cancer Institute, 2018. https://www.cancer.gov/about-nci/organization/crs/research-initiatives/team-science-field-guide/collaboration-team-science-guide.pdf. Accessed on September 25, 2022.

6 Brian Mathews, et al. 2023. "Multimodal Archives: A Toolkit for Collecting Robotics and Other Complex Material in a Research Ecosystem." Carnegie Mellon University. doi:10.1184/R1/22232692.v1, 27.

7 Mathews et al., 11.

8 Mathews et al., 41.

Bibliography

Bennett, L. Michelle, Howard Gadlin, and Christophe Marchand. *Collaboration Team Science: Field Guide* [Rockville, Md.]: U.S. Department of Health & Human Services, National Institutes of Health, National Cancer Institute, 2018. https://www.cancer.gov/about-nci/organization/crs/research-initiatives/team-science-field-guide/collaboration-team-science-guide.pdf. Accessed on September 25, 2022.

Conway, Martha O'Hara, and Merrilee Proffitt. 2011. *Taking Stock and Making Hay: Archival Collections Assessment.* Dublin, OH: OCLC Research. https://doi.org/10.25333/C33S6M.

Mathews, Brian, Katherine Barbera, Kathleen Donahoe, and Lauren Herckis. "Multimodal Archives: A Toolkit for Collecting Robotics and Other Complex Material in a Research Ecosystem." Carnegie Mellon University, March 8, 2023. https://doi.org/10.1184/R1/22232692.v1.

McCorduck, Pamela. *Machines Who Think: A Personal Inquiry into the History and Prospects of Artificial Intelligence.* Natick, MA: A.K. Peters, Ltd. 2004.

Newell, Allen. Transparencies (Hard Copies) -- Meeting with Edith Martin on the Software Engineering Institute - Graphic Material, 19 December 1983. Allen Newell Papers. Carnegie Mellon University Archives, Carnegie Mellon University.

Robotics Institute, Carnegie Mellon University. "Research Overview." Accessed December 16, 2024. https://www.ri.cmu.edu/research-overview/.

10

Building an Inclusive Science Archive in India[1]

Deepika S and Anjali Ramachandran

Abstract

This paper is about strategies implemented in a science archive in India to include more material from marginalized groups in science, particularly from women. We discuss our attempts to rethink our collection mandate, and why we have had limited success with the volume and diversity of material that gets accessioned in the archive. We also detail the challenges of sustaining a public archive in India and the fundamental tensions between building an inclusive archive and preserving scientific heritage with limited resources. We examine archival theory and its limitations in our context – especially the principles of provenance and original order, and the concept of archival integrity – along with problems of digital access. In doing this, we rely on feminist methodologies such as feminist standpoint appraisal to re-frame our approach to archiving, shifting our focus from the archive as an institution to include the people and communities that are archived and involved in the process of archiving.

"Where is the Veronica Collection?"

This question was put to the Archives at NCBS[2] at a gathering in 2023 to commemorate the life and work of Veronica Rodrigues, a molecular biologist – known for pioneering work on studying how information about smell and taste is coded in the brain – who passed away at the age of 57 in 2010. The aim of the gathering, organized by Gita Chadha (the Obaid Siddiqi Chair[3] at the Archives for 2023-2024), was to kickstart a conversation about Rodrigues' life as a scientist with the aim of answering wider questions about what it means to be a woman in science in India. As a sociologist and feminist science studies scholar, Chadha was interested in Rodrigues because she saw her as a complicated figure. Rodrigues was a Kenya-born microbiologist who came to India, while stateless, to work with Obaid

Siddiqi, a neurogeneticist at the Tata Institute for Fundamental Research (VijayRaghavan and Bate 2010, 1493). She was also part of the core group of senior scientists who later set up the National Centre for Biological Sciences (NCBS). Rodrigues is remembered as a pioneering woman scientist who navigated the labyrinth of an elite scientific institution in India.

At the gathering, some of Rodrigues' former students and colleagues shared their memories of her, wondered how to gather her personal papers, and spoke about preserving her legacy. She was a single woman without children, regarded by some as a perfectionist who did excellent work, by some as a harsh and exacting supervisor, while others remembered her warmly. The Archives is home to a small number of documents created by or about Rodrigues, and these are illustrative of critical issues related to migration, citizenship, and gender in science. (Figure 1) They are scattered across two NCBS institutional collections as well as in the personal papers of two of her colleagues at NCBS.[4] Digital access to some of these documents has also led to increased public interest in Rodrigues.

Chadha (in her gently provocative manner, and to the initial disquiet of the Archives team) asked why there is no Veronica Rodrigues collection at the Archives – a question that has been posed a few times since its very inception in 2019. In response to Chadha's question, the Archives team was slightly defensive; team members explained that archivists do not curate material, and cannot remove documents from their context of creation to reveal their diverging, layered meanings. The Archives does not have a Veronica Rodrigues collection simply because it did not receive any material from Rodrigues or her legal heirs. Her material is part of other collections, and the Archives wanted to honor provenance and register the origins of the collections.

But the question itself helped the authors of this paper (we'll refer to ourselves as Anjali and Deepika) realize that the very archival framework we were working within was 'hiding' (Banks 2023) marginalized voices in our conventional science archive. This unsettled us, because we considered ourselves sensitive to the problem of diversity in our collection mandate. We had both joined the Archives in late 2022, broadly around the time that

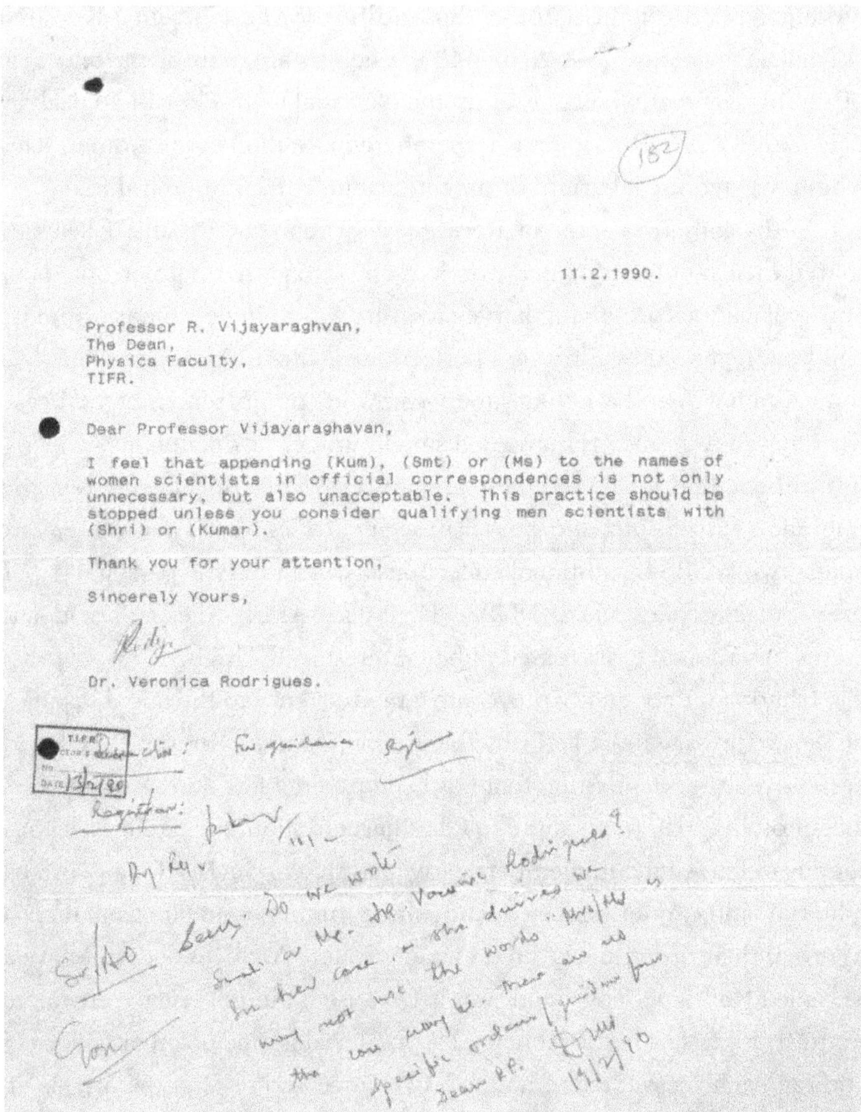

11.2.1990.

Professor R. Vijayaraghvan,
The Dean,
Physics Faculty,
TIFR.

Dear Professor Vijayaraghavan,

I feel that appending (Kum), (Smt) or (Ms) to the names of women scientists in official correspondences is not only unnecessary, but also unacceptable. This practice should be stopped unless you consider qualifying men scientists with (Shri) or (Kumar).

Thank you for your attention,

Sincerely Yours,

Dr. Veronica Rodrigues.

Figure 1. NI-00-1-2-1-9, Letter from Veronica Rodrigues regarding the use of personal titles for women scientists in official correspondence, NCBS Establishment Office Collection, Archives at NCBS

the Archives received a grant to build an open access digital archive for the history of science in contemporary India, and its staffing had expanded to include several women, including us. We had both trained in humanities, and had made a mid-career switch to archiving, learning on the job, and had worked in archives at other science institutions beforehand. Like most 'non-science' staff at the peripheries of elite scientific institutions, our lived experience is affected and informed by the undertow of hierarchy, politics and prejudices entrenched in these spaces. In conversations over tea and lunch breaks – and as our friendships with each other and within the team cemented – we began to air our frustration at being women whose professional labor led to the archiving of only men. When we joined, there were eleven collections at the Archives, all either related to NCBS institutional material or from male scientists. We wanted to change this and had already had several conversations with other supportive colleagues about it. We had taken a few steps towards (what we thought was) addressing the problem but found that we were still only bringing in material from male – and largely upper caste – scientists. Why did we find ourselves failing miserably at archiving more women in the sciences? And even when we did have some archival material from women, like in Rodrigues's case, why did it seem so difficult to find in the archives? These questions forced us to examine our own professional practices as archivists, as well as archival theory, to find answers.[5]

We agreed that we wanted to build an archive that reflected the way science is practiced in India, with archival material from people of different castes, religions, regions and genders involved in various kinds of work within science. The cost of not doing so seemed too high: science in India is often seen as being the purview of upper caste men – who are perceived to be "natural inheritors of scientific practice" (Thomas 2020, 1). Religious, sacerdotal practices become conflated with culture and are treated as compatible with – and especially conducive to the study of – science. Ideas of merit and scientific objectivity are calibrated to make invisible hierarchies of caste, class and gender (Thomas 2021). These ideas also erase, exclude and discriminate against marginalized groups (Gupta 2020) in scientific institutions in India – and this has consequences.[6] If we didn't change the kinds of material we were collecting in our archive, were we not complicit in propagating a mostly male and Brahmanical hegemony in science? Reckoning with the absence of a Veronica Rodrigues collection at the Archives became in many ways the starting point for deeper conversations amongst us and our colleagues about building an inclusive archive.

Prioritizing diverse material

Our initial efforts to develop an archive that attempted to represent scientific culture in India, and not just build an archive of upper caste men in science, included modifying our accessions workflow, creating a system of points to help us decide which donors we should be prioritizing while collecting material. On the list of potential living donors or family members or potential organizations that we could reach out to for material, we created additional columns against their names to add points for age, gender, caste, religion, geographical region, along with a column provisionally titled 'non-mainstream' (largely to look beyond people considered 'notable' or 'eminent', and as an umbrella to cover scientific research that seemed important but was carried out in an unconventional discipline or institution or something else that we weren't able to foresee). Adding up these points helped us prioritize material that we wanted to pursue to help us fulfil our goal of having diverse material. This did not stem the influx of material from upper caste male scientists; the Archives continued to be approached by many and accessioned some of their papers.

We also spoke to one of our funders about our annual indicative targets of the number of digital objects we make available for public access. We explained that we had noticed that archival material from marginalized groups tended to be smaller in volume and harder to come by. We didn't want to overlook this material while working towards our targets, but we realized early on that it would take more resources to pursue diverse material. In August 2023, we decided to pause accessions from male scientists beginning in February 2024 (by then, we had processed 18 new collections, two of which were collections of partnered individuals that included women, and only one of which was a standalone collection from a woman).

This focus on women seemed necessary to us for several reasons. In our experience, acquiring someone's papers, irrespective of their gender and social location, is not easy: India doesn't have a robust culture of preserving scientific heritage (Chowdhury 2013, 372), and we find a lot of our effort directed towards simply making a case for the archive. In this context, where it isn't a given that archival material – even from famous men – will find their way into an archive, we also feel the enormous burden of preserving scientific heritage as one of very few science archives in India. Limiting accessions from male scientists has not been an easy decision to make. Finding diverse voices and bringing them into the archives has been a difficult task, especially when you consider the realities of working in an archive where adequate funding and staffing cannot be taken for granted, and the daily pressures of cataloguing, preservation and providing researcher access can overwhelm the desire to step back and look at the larger picture. Coupled with this, as mentioned before, we found that material from women in India – even 'notable' and 'eminent' ones – was harder to collect.[7] The papers of the well-known botanist E. K. Janaki Ammal, for instance, are lost or destroyed, and most of what is known about her work comes from the papers of her mentor and collaborator, C. D. Darlington. (Damodaran 2017, 832)

In our interactions with donors, we noticed that male scientists tend to have created more written records and are familiar with archives and more open to donating their material to the archives; we have rarely had to convince a male scientist of their place in history. We have also observed that many women did not see themselves as belonging in our archive – we've heard women wonder if their material is valuable enough to be deposited in an archive, and we've had one potential donor decline to give us material after initially agreeing to it because they felt they had been sidelined and diminished over the course of their career and there was no point to preserving their material. The lack of archival material from women in our own archive also likely acts as a deterrent (and we hope that the converse will be true: as we've seen in other contexts, like sometimes attracts like, and that's our jokey explanation within the team for why we have so many collections related to ornithology).

In addition, a significant amount of the material we have attempted to collect from marginalized groups has tended to be more contemporary in nature, largely from the 1990s onwards, which poses issues of ethics and sensitivity. The Archives usually only uploads records older than 30 years.[8] Since the Archives' funding depends on collecting and making available free digital access to all our material, having such contemporary material is tricky, and has led us to develop a sensitivity matrix for making oral histories available, though we are yet to develop one as detailed for physical collections. Even without the constraints of time, we have seen that material from marginalized groups tends to contain sensitive and private information, sometimes involving trauma, violence, or political persecution[9] and how to process and make this available is something we must work harder at reckoning with.

To expand our understanding of science in India, we began to look beyond the confines of leading scientific institutions to community and citizen-led efforts in popularizing and instrumentalizing science in issues related to the environment, indigenous rights, reproductive rights and social movements. Informed by feminist science studies' critiques of the gendered nature of scientific practice and particular disciplines within science, including a hierarchy of applied versus fundamental scientific research (Chadha and Achuthan 2017, 33–36), we reached out to activist and women's movement networks in India to document scientists, educators, organizations, activists, artists and writers who have worked on feminist issues around science research and policy, public health, and women's rights to their bodies. We also looked at scientific fields that were more likely to have more women, such as medicine and public health, food technology, and agricultural science.

We've had limited success: by December 2024, after nearly a year of pausing accessions from men, the Archives brought in 11 collections[10] from women (though these do not compare in volume or in nature to the collections that we have from men), two from environmental organizations, and a few more from male activists whose papers are related to citizen-led conservation efforts. This re-enforced for us the value in looking beyond stereotypical representation of women in STEM institutions as scientists, and validated our broader approach to collecting material beyond traditional STEM fields.[11] All of these efforts also confirmed for us that the Archives, in the course of 'business as usual', slips naturally into reflecting the Brahmanical and patriarchal structure of science (Chadha and Achuthan 2017, 33–36), and that avoiding this will require significant structural changes. For meaningful change, we know that our efforts have to go deeper, to address problems beyond the culture of science, to examine the framework of the Archives itself.

Feminist standpoint theory and the archive

In late 2023, Gita Chadha taught a course at NCBS introducing students of science to feminist science studies,[12] during which she discussed the philosophical lineage of feminist science studies as well as feminist methodological tools such as principles of care, intersectionality, a focus on lived experience, reflexivity, and feminist standpoint theory. The authors of this paper had attended the lectures, and though we were both familiar with feminist theory, the course helped crystallize some of our questions about archiving, and reminded us of the influence that feminist thought has had on archiving, even if some of these influences are not always made explicit.[13] It pushed us to research feminist standpoint epistemology and other tools that we could potentially deploy in our own work. That led us to Michelle Caswell's paper on feminist standpoint appraisal (Caswell 2021a), which was the diving board off which we sprung into feminist archival theory.

Feminist standpoint theory reminds us that knowledge is not value-neutral but situated; that people who occupy positions at the margins may have a different perspective that is valuable in overcoming this supposed neutrality, and that examining power relations should begin with the lived experience of those on the margins. (Harding 2004) Caswell applies this to archives, calling attention to the role of the archivist "as a socially located, culturally situated agent who centers ways of being and knowing from the margins." (Caswell 2021a, 2) While subjectivity, both in relation to the creation of archival records as well as the position of the archivist, is by no means a new idea – even influential archival theorists discuss this in their work (Ketelaar 2001; Schwarz and Cook 2002) – Caswell makes the case for embracing this subjectivity, rather than working to overcome or neutralize it. Caswell also highlights using the "view from the margins" in the determining archival value and posits that feminist standpoint appraisal "acknowledges all knowledge is partial and attempts to diminish the perversity of that partiality by serving those most in need." (Caswell 2021a, 23) It also forces us to ask crucial questions of ourselves: who are we most accountable to as archivists? Who is our work really for? How do we calculate who is "most in need"?

Feminist standpoint appraisal helped us understand our own work as archivists – it helped reinforce not just the importance of having diverse archival collections but also the importance of understanding how the identities of the people[14] who work in the archive enable different aspects of our work. It encouraged us to bring our whole selves to work, including our personal and political values, and clarified that our own identities and experiences of gender and caste could prove useful in navigating the work we had set out for ourselves. It also helped us listen to our instincts. One of the tensions we have struggled with has been the need to be "objective" and "impartial" as archivists. The International Council on Archives' Code of Ethics states under its very first statement about protecting the integrity of archival material: "The objectivity and impartiality of archivists is the measure of their professionalism." (ICA, 1996) In our practical experience as archivists trying to archive the lives of women scientists, being impartial to people of all genders has not meant equal efforts in all directions, but rather a particular and additional emphasis on people and organizations from marginalized groups for meaningful effect. And if being objective requires not being influenced by one's personal feelings in the carrying out of one's duty (a superhuman ask of anyone, especially archivists – who in our experience are largely young, female, working on inadequate salaries, and are fairly low in the pecking order at a science institution), then we deny the motivations and decisions that drive the work we do to bring in diverse material.

We find that as archivists, our lived experience, our politics, and our positionality are inseparable from – and of vital importance to – the work that we do. Understanding science in contemporary India demands not just an understanding of technical knowledge but also of cultures in science and hierarchies in science – of disciplines, of gender, of caste, of class, of disability, and other axes – and a trauma-informed approach to archiving (how else do we adequately process the papers of women in our archive that pertain to domestic violence, physical violence at the workplace, or conduct oral history interviews of women who speak of sexual assault?).[15] Remaining attentive to discussions and debates around these are crucial in order to strive to archive material related to science in contemporary India with nuance, honesty, fairness and sensitivity. This also requires that we see ourselves as "caregivers, bound to records creators, subjects, users, and communities through a web of mutual affective responsibility." (Caswell and Cifor 2016, 23) Feminist standpoint appraisal encourages us to leverage aspects of our positionality while acknowledging and working on our blind spots (and not simply relying on our identities), even as it forces us to think about what we collect at the Archives and why. After all, archival appraisal is "*the* critical archival act by archivists", the "gateway function to all subsequent archival activity", the archivist's "first responsibility." (Cook 2009, xiii–xiv)

We don't see it as a coincidence that our first standalone collection from a woman scientist – the Krishnaja AP Papers – entered the archive after the team had expanded significantly to include more women archivists, some of us with an understanding of science, feminist theory, women's movement and activism, and access to networks of researchers working on women in science. This first collection from a woman arrived at the Archives through a collaboration with Nandita Jayaraj, a science communicator, author and one of the founders of The Life of Science, a feminist science media project. Jayaraj did a series of oral history interviews with six scientists who were women (and prior contacts of hers), nearly all of whom have subsequently donated or expressed an interest in donating their papers to the Archives at NCBS. Deepika was also present in preliminary meetings with all of the interviewees and stayed in contact with them even after the collaboration with Jayaraj ended. The success of this project convinced us of the need to have diverse staff and collaborators with varied personal networks, the need to build and maintain relationships with potential donors, and of the efficacy as a strategy of using oral history interviews to build these relationships (which in several cases also led to the acquiring of physical collections).[16]

In our experience, having diverse staff has played an important role not just in reaching out to a range of potential donors, but also in co-creating archival records such as oral histories. They also brought to the archives their networks, politics and interests. It was through the work, interests and connections of Dr. Samira Agnihotri, an ecologist who studies birdsong in the Biligiri Rangan Hills (better known as the BR Hills) and a former colleague at the Archives, that the first collection from a people's movement – the Wayanad Prakrithi Samrakshana Samithi Papers – eventually entered the Archives.[17] Agnihotri was also was interested in the papers of Leslie Coleman (a Canadian plant pathologist and entomologist who served as Director of the Department of Agriculture in the erstwhile Mysore State from 1913 to 1934) in the Archives as it included several photographs and family albums that record images of forests in the BR Hills, cultural and religious practices in Mysore, and hunting, among other things. Created as it was during colonial rule in India, the collection also contains ethnographic portraits of people from various tribes, as well as other material that may be seen as racist and offensive. Amongst these images are photographs of people from the indigenous Solega community that live in the BR Hills. Agnihotri was able to inform some members of the community about the images in the collection and show it to them in person. Through Agnihotri, who has strong personal networks in the region, we have built a relationship with a few younger members of the community who have also visited the Archives and conveyed changes that the community would like to add to create reparative descriptions for the Coleman photographs of BR Hills and the Solega.

Moreover, we actively shape the kinds of physical material we receive through our interactions with donors and the decisions we make about what to collect. When packing material of possible archival value at the home of Krishnaja AP, a retired cytogeneticist in her 70s, Anjali decided to include a folder of unusual material along with papers about Krishnaja's life and career: clippings of food recipes from a range of women's magazines. At first glance, it seemed to say little about Krishnaja's work as a scientist, but Anjali included these recipes because in conversation with Krishnaja at her home in Mumbai, she understood that they seemed important to Krishnaja and had been preserved in the same way and in the same location as all of Krishnaja's other material about her career in science. Once Krishnaja's material was transported to the archives and an appraisal of the material was carried out, Anjali decided to retain the recipes with the collection. Deepika was initially concerned that retaining them might reinforce stereotypes about women scientists (considering this was the Archives' first standalone collection from a woman, and considering the accomplishments of women scientists are sometimes eclipsed by their domestic roles – like in the *New York Times* obituary for Yvonne Brill, an American rocket scientist, that originally opened by listing her domestic achievements, including making a "mean beef stroganoff"[18]). But Krishnaja's recipes (Figure 2) now form a small part of her collection, listed as a single folder in Sub-series 1 of Series 3: Personal & Hobbies; the first two series in the collection being Correspondence and Research. The recipes have no equivalent in any of the collections at the Archives that include papers from male scientists and reveal the gendered nature of domestic duties and how when archiving the lives of women scientists, there may not be a neat line between the personal and professional.

We have also learned much from queer perspectives on archives. Alana Kumbier's writing on ephemeral material (Kumbier 2014) had us nodding emphatically at several of the points they raise about the challenges of archiving queer lives that resonated in our work. Kumbier draws attention to the value of marginal or ephemeral materials documenting queer lives, as well as their affective states. They also point to the difficulty of accessioning material from people outside traditional familial networks and in the absence of proper channels, the difficulty of making material acquired through 'extra-legal' channels with researchers. They write of 'haunting' by archival absences – and that this can be a productive force. Mainstream archives can learn much from feminist and queer collectives who, as Mela Dávila-Freire points out, "have been remarkably active in their endeavors for archival emancipation, not only building their own archives and pointing out historical absences or gaps in hegemonic documentary collections, but also identifying information categories that mainstream archives have not

Figure 2. MS-028-3-1-4-15, Advertisement in a women's magazine, Krishnaja A P Papers, Archives at NCBS

explicitly recorded – such as affects, feeling or emotions." (Dávila-Freire 2020, 94)

Kumbier presses "the idea that we can develop situated, responsive archival practices in collaboration with specific constituencies" (Kumbier 2014, 32) and that is what we aim to work towards. Here, we should say that we are acutely aware that the Archives is not a feminist or queer or community archive. It is situated not in the margins, but firmly in the center. It collects across genders, disciplines, and kinds of institutions, receives corporate and charitable funding, and has material from privileged groups as well as oppressed ones. The large collections in the Archives from 'eminent' male scientists bring visibility and funding, and enhance its reputation as a 'serious' archive. But we do find ourselves haunted by the archival absences of marginalised voices, and we wonder if a way to address this (without appropriating or assimilating these voices) might also be to use the tools we've been armed with by feminist, queer and community archives and apply them to all the collections we have – including those of upper caste men.

Can we not harness a little of that "archival emancipation" for the voices in our archive?

Reframing a mainstream science archive

Critical feminist perspectives on archives help us listen to our instincts, even when they aren't always in alignment with archival theory. And they give us a clarity of vision, especially when the language of feminism and social justice is pervasive in daily life, and organizations frequently deploy this language (think "intersectional" or "DEI") as a substitute for meaningful action. It centers not the archive as an institution, but the people whose lives are archived and the people doing the archiving; the hierarchies, hegemonies, microaggressions, violence and politics that this involves, as well as the joys, friendships, resistance, solidarities and "structural care" (Caswell and Cifor 2021), that comes with it. Critical feminist perspectives reframe the responsibility of archivists as being to people and communities, rather than to the archive or its parent institution[19], or even to posterity.[20] It reminds us to consider archival principles not as universal, but as ideas that can be extended, distended, continually questioned and occasionally buckled, not just when applied to collections from marginalized groups, but for all our collections.

As we've outlined, having material from marginalized groups in the archives requires expanding our skills, theoretical frameworks, and our practice as archivists. It also exerts a force on the other material in the archives and changes the very texture of our archive – it queers, feminizes, diversifies, and politicizes it. Krishnaja's magazine cuttings, with photographs of housewives contributing dessert recipes and advertisements with models in swimsuits selling hair removal cream, poke at our notions of what a 'serious' archive is. Grappling with all of these tensions requires rethinking and, in some instances, unlearning our training as archivists, and we have found feminist archival theory particularly useful in translating ideas into practice.

Archival processing and MPLP

We know that detailed processing and archival description is important to make marginalised people visible in the archives. Kumbier writes, "If LGBTQ materials are in archives and libraries, but are inaccessible due to inadequate or censorious description, they cannot enter the historical record." (Kumbier 2014, 29) At the Archives, we only create folder level descriptions; item level description could help give a richer sense of archival collections. While "MPLP" – More Product, Less Process (Greene and Meissner 2005) seems a seductive strategy as it addresses the archivist's workload, it especially requires that we pay careful attention to the people and communities left out by 'efficiency'. Insufficient care when cataloguing or incomplete archival description can render important material invisible. An example: Mina Swaminathan is an educator, activist and public figure whose presence in the Archives is primarily through photographs from the collection of her spouse, the agricultural scientist, MS Swaminathan. Her voice is represented in a personal diary mislabeled by the Archives as belonging to Swaminathan, in which she recorded her pregnancies in 'birth reports', the early childhood of her three children, and a set of Tamil lullabies. Another small set of papers in the same folder comprises another 'birth report' penned by one of her daughters. The folder is simply named 'Personal Diary - 1959', accompanied by the description: 'Personal diary maintained by M.S. Swaminathan for the year 1959.' The Archives has since edited the finding aid, with a note about the original description and the revisions made to that. Revisiting archival descriptions for the MS Swaminathan Papers may yield more traces of Mina Swaminathan and others. We also hope to shortly begin an appraisal and metadata project (accessing varied controlled vocabularies beyond only Library of Congress terms[21]) to uncover gaps and exclusions in the papers.

The MS Swaminathan Papers is the largest collection in the Archives, documenting the history of farming, agricultural technology and governance in post-independent India through Swaminathan's personal material. It also includes institutional material related to the Indian Agricultural Research Institute, the Indian Council of Agricultural Research, the MS Swaminathan Research Foundation, and their collaborations with several international organizations. It is an extraordinary collection simply because of the variety of material within it and as it is presently catalogued, it does not respond to the multifarious voices, identities, politics and policies that shaped the narratives within the collection, and perhaps misrepresents provenance in terms of Swaminathan being its sole originator. The very naming of this collection perhaps unwittingly props up the idea of the 'great man of science', and revisiting this naming might be a good starting point to re-examine the collection.

The idea of renaming this collection originated after the naming of women's collections at the Archives. Of the 11 collections from women processed at the Archives between February and December 2024, one is a set of just six photographs, while another has only two academic theses. We would typically arrange these under a larger collection called "Small Donations" at the archives, but we ultimately decided that having these as separate collections named for the women who were their originators was a way of making space for them in the Archives. Elizabeth Yakel writes:

"Through the process of selection of information for inclusion and choice of access points, archivists reveal and conceal, making finding aids political statements" (Yakel 2003, 20). Our insistence on having separate named collections for women and not burying them in finding aids is no less political than the decision to not prioritize their visibility in the first place. Moving forward, the Archives team made a decision to scrap Small Donations as a collection entirely.

Collecting material related to women has enabled us to critically examine our existing collection and re-fashion our description standards in order to align with and respond to feminist and queer methodology. We aim to tackle the problem of identities rendered invisible and narratives while also, we hope, not limiting ourselves to the "search-and-rescue model, or savage-to-salvage model, where the lost histories of the past [are] recuperated and reinstated within more liberatory histories of the present" that Anjali Arondekar diagnoses (Arondekar, Cvetkovich, Handhardt, et al 2015, 215).

Archival Integrity, Provenance and Original Order

The ICA's code of ethics states: "The primary duty of archivists is to maintain the integrity of the records in their care and custody" (ICA 1996). We struggle with this on occasion – in the past, the Archives has gone to enormous lengths in order to maintain archival integrity, including de-accessioning and transporting material donated to us by a male scientist so that it could reside with his other material in an archive across the country. But in one ongoing instance, when a well-known woman scientist passed away during the process of donating her material to the archives, we made a decision that we are still working through. Once it was clear we may not have access to all of her material that currently resides at her workplace, we began to negotiate with the scientist's institution to keep at least a portion of her material – out of concern that what remains at the institution may not be preserved, to essentially salvage what we can. This would mean we do not preserve either archival integrity or original order, in this instance.

The classical archival principle of respect des fonds, or original order, for instance, is still held up as a standard, despite criticism over the years of its applicability to personal papers. In 1982, Frank Boles pointed out the hesitation by archivists to "tamper" with original order as "archival shibboleths" will be disturbed (Boles 1982, 30).[22] One could perhaps say the same of archivists today, ourselves included. Some archivists have complicated the idea of what we might consider 'original' order in the first place, and Heather MacNeil cautions that it is "simply one of many possible orders a body of records will have over time and, therefore, its privileged status needs to be reconsidered" (MacNeil 2008, 24).

The hiding of women in archival collections has been the subject of reparative projects that curate archival material or tell stories from the archive to make these figures more visible. The project 'Visualising Women in Science' using the collections of the American Philosophical Society[23] is a recent example of this. To answer the Veronica question that we raised at the start of this paper, we initially considered a similar project of digital curation to bring her papers together. However, we worry that this still relies on a reparative project as the first instance in which the user of the archive discovers her. What if we were to disregard original order, to create a sub-series within various institutional collections named for Veronica Rodrigues? Or what if we were to create a new collection entirely, adequately cross-linked to other collections so that its context is not lost? We've received firm pushback from a colleague for suggesting this (with valid questions: do we do this for all people who are marginalized? How do we decide who falls within this category? Who might we exclude while doing this? How can we possibly do this at scale?)

We don't have an elegant answer here, just frustration that abiding by the principles of provenance and original order continues to faithfully reproduce the inequities in our culture. But perspectives from feminist, queer and community archives remind us that diverse material may require varied and perhaps unconventional approaches. They may at times call for queer revisionism or imagined history (Kumbier 2014, 44), or provenancial fabulation (Lapp 2023) too.

Digital Access

And finally, we look beyond our own daily practices at the archives to consider the larger system in which we operate. The existing funding ecosystem at the Archives, primarily from the global north, incentivizes digital access as an intrinsic part of archival processing. As newer digital archiving projects mushroom in India[24], more and more South Asian histories and narratives are becoming accessible to users across the world, whereas traditionally well-funded and elite archives across the Global North (also containing records about South Asia) have yet to provide open digital access to most of their holdings. In this tethering of open digital access to the continued patronage of archiving projects, we see ourselves reproducing unfair labor relations in the field of archiving. The reality is that labor is cheaper in countries like India, and though digitization requires skilled technicians, there is a prevalent casualization of this labor in such projects (which our colleague refers to as the consequence of "digitization on a dime").

As an archive, we have to consider whether we are propagating the very inequities that we are trying to make visible in our collections. These are uncomfortable questions to confront as archivists, but we need to be self-reflexive about the practices and processes in our institutions. With funding for digital archives, we are able to exponentially expand the scope and ambition of our work, sometimes benefitting from pre-existing caste-bound unfair labor conditions. We must therefore be careful to create policies that encourage fair labor practices, equitable pay and a long-term commitment to skilling, to ensure that commitment to the archive also extends to all the people who make up the archive.

Though digital access to material can democratize access, we wonder if the delivery of the digital material in its present form as digital surrogates in an archival catalogue makes them difficult to navigate for users unfamiliar with archival research. We cannot take primary source literacy for granted in India, and worry (that thorny question again: who is our work for?) that our material is largely only accessible to researchers from Western universities who have the funding, time, social and cultural capital to decode and access archives. For stories from marginalized and underrepresented groups to be truly accessible to the same communities and people who had created and have preserved their stories, access at the archives has to be transposed and extended beyond archival catalogues. Community outreach needs to enfold the creators and custodians of diverse material, instituting them as long-term partners to transform the way the material is made accessible.

At the Archives, this has meant connecting researchers to the donors and previous custodians of the material whenever this is possible. Otherwise an inclusive collection mandate remains a job half done; the diverse stories in the archive will require a researcher to 'discover' it and then retell the story, often as an impersonal encounter with the material, instead of framing their dialogues with creators, their stories and archivists into the research output. We are also working towards building a multi-lingual catalogue for one of our community collections so that the archival catalogue can be made accessible to the creators of the material. The Archives has recently begun a research fellowship program to enable researchers from India to conduct extended archival research – some of whom have been first-time users of the archive – which we hope will enable access to local users beyond the confines of academia.

Digital access to material from marginalized groups can highlight and legitimize such material and their positioning alongside other collections in the archive can help frame the narratives within the archive. However, this is not without problems – in a case study of three feminist archives and their approach to digitization (Eichhorn 2014, 227-37) highlights the projects' response to the materiality, context of creation and public perception of the collections, which compelled them to decide not to digitize the material. At the Archives, digital access also poses serious ethical questions – it is imperative that we create safe spaces for material from the margins, whether it is contemporary material from anti-government protest movements, civil rights activists or material containing evidence of violence, abuse and disenfranchisement. We need to reconcile the democratizing principle of digital access with its potential to be weaponized by bad actors and entities. To respond to this, limiting access in the short term and providing layered access in some instances (as opposed to blanket digital access which also enables trawling by AI) might be a way to ensure their long-term preservation at the archives. We hope to learn more from introducing a participatory approach to archiving and digitizing material from marginalized groups,[25] to enhance our current process and inform our decisions as archivists.

Conclusion

In India, professional training in archiving is hard to come by; few universities offer degrees in archiving, while a handful offer short courses and diplomas (Anjali has a diploma in archiving, while Deepika has no formal training). This has been both challenging and useful; while our understanding of archival theory had to be developed independently, we've found it easier to look beyond canonical archival theory when approaching a problem like the absence of a Veronica Rodrigues collection in the Archives. We learned quickly that good intentions and collection mandates weren't sufficient to bring in diverse material, and that structural changes were needed instead. In the year spent focused on attempting to halt 'business as usual' and search for better strategies to include marginalized groups in science, we have learned immensely from the writings of archivists working in feminist, queer and community archives, from archivists grappling with their identities, subjectivities and politics within the archival profession, and from contemporary theorists searching for ways to forge through the challenges of archiving. As we've outlined above, we have found feminist methodologies particularly valuable in weaving the theories into daily practices and processes at the Archives. This has helped us envision a roadmap for the Archives to develop as an inclusive space that is reflective of the culture and practice of science in India, while also recognizing the decisions of those who choose not to participate in the project of archiving. We'd like the Archives to be a place where the memories of different kinds of people are preserved, not to assimilate or flatten them to serve convenient narratives about science in India, but to make space for their unique contexts and particularities. We may never collect and preserve the Veronica Rodrigues Papers, but working towards building diverse collections has helped center the Archives around different stakeholders as important agents in the community of the archive – and shape the Archives to be a home for their material.

Bibliography

Archives at the National Centre for Biological Sciences (NCBS), India, in collaboration with the Milli Archives Collective. 2023. *Archives, Ethics and the Law in India: A Guidebook for Archivists in India*. https://ethics--law.archives.ncbs.res.in/

Arondekar, Anjali, Ann Cvetkovich, Christina B Handhardt, et al. 2015. "Queering archives: A roundtable discussion". *Radical History Review*. 2015 (122): 211-232. https://doi.org/10.1215/01636545-2849630.

Banks, Catherine. 2023. "Decoding Hidden Women: Feminist digitisation practices in the Tale Archive". Tale Archive Blog, October 23. https://blogs.ed.ac.uk/garg/2023/10/23/decoding-hidden-women--feminist-digitisation-practices-in-the-tale-archive/

Boles, Frank. 1982. "Disrespecting Original Order." *The American Archivist* 45 (1): 26–32. http://www.jstor.org/stable/40292457.

Caswell, Michelle and Cifor, Marika. 2016. "From Human Rights to Feminist Ethics: Radical Empathy in the Archives." *Archivaria.* 81 (May): 23–43. https://archivaria.ca/index.php/archivaria/article/view/13557.

Caswell, Michelle. 2021a. "Dusting for Fingerprints: Introducing Feminist Standpoint Appraisal." In *Radical Empathy in Archival Practice*. Edited by Elvia Arroyo-Ramirez, Jasmine Jones, Shannon O'Neill, and Holly Smith. Special issue, *Journal of Critical Library and Information Studies*. 3(2). https://doi.org/10.24242/jclis.v3i2.113.

Caswell, Michelle. 2021b. *Urgent Archives: Enacting Liberatory Memory Work.* Routledge.

Caswell, Michelle and Marika Cifor. 2021. "Revisiting a Feminist Ethics of Care in Archives: An Introductory Note." *Journal of Critical Library and Information Studies*. 3 (2). https://doi.org/10.24242/jclis.v3i2.162.

Chadha, Gita, and Asha Achuthan. 2017. "Feminist Science Studies: Intersectional Narratives of Persons in Gender-Marginal Locations in Science." *Economic and Political Weekly* 52 (17): 33–36. http://www.jstor.org/stable/26695963.

Chowdhury, Indira. 2013. "A historian among scientists: Reflections on archiving the history of science in postcolonial India." *Isis* 104 (2): 371–380.

Christen, Kimberly and Jane Anderson. 2019. "Toward slow archives." *Archival Science*. 19: 87-116. https://doi.org/10.1007/s10502-019-09307-x.

Cifor, Marika. 2016. "Affecting relations: introducing affect theory to archival discourse." *Archival Science*. 16**:** 7–31. https://doi.org/10.1007/s10502-015-9261-5.

Cifor, Marika and Stacey Wood. "Critical Feminism in the Archives." In "Critical Archival Studies." Edited by Michelle Caswell, Ricardo Punzalan, and T-Kay Sangwand. Special issue, *Journal of Critical Library and Information Studies*. 1 (2). https://doi.org/10.24242/jclis.v1i2.27.

Cook, Terry. 2009. "Foreword." In John Ridener. *From Polders to Postmodernism: A Concise History of Archival Theory.* Litwin Books.

Cook, Terry. 2013. "Evidence, memory, identity, and community: four shifting archival paradigms." *Archival Science* 13: 95–120. https://doi.org/10.1007/s10502-012-9180-7.

Damodaran, V. 2017. "Janaki Ammal, C. D. Darlington and J. B. S. Haldane: Scientific encounters at the end of empire." *Journal of Genetics* 96: 827–836. https://doi.org/10.1007/s12041-017-0844-1.

Dávila-Freire, Mela. 2020. "Reading the archive against the grain: Power relations, affective affinities and subjectivity in the *documenta* Archive" *Art Libraries Journal*. 45 (3): 94-99. https://doi.org/10.1017/alj.2020.14.

Drake, Jarrett M. 2017. "I'm Leaving the Archival Profession: It's Better This Way." *Medium* https://medium.com/on-archivy/im-leaving--the-archival-profession-it-s-better-this-way-ed631c6d72fe.

Eichhorn, Kate. 2014. "Beyond digitisation: a case study of three contemporary feminist collections." *Archives and Manuscripts*. 42 (3): 227-237. https://doi.org/10.1080/01576895.2014.958866.

Greene, Mark A., and Dennis Meissner. 2005. "More Product, Less Process: Revamping Traditional Archival Processing." *American Archivist*. 68 (2): 208–263. https://doi.org/10.17723/aarc.68.2.c741823776k65863.

Gupta, Namrata. 2020. "Rationalizing gender inequality at scientific research organizations: A reproduction of the Indian socio-cultural context", *Equality, Diversity and Inclusion*. 39 (6): 689–706. https://doi.org/10.1108/edi-09-2018-0168.

Harding, Sandra. 2004. "Introduction: Standpoint Theory as a Site of Political, Philosophic, and Scientific Debate." In *The Feminist Standpoint Theory Reader: Intellectual and Political Controversies.* Edited by Sandra Harding. Routledge.

International Council on Archives. 1996. *Code of Ethics.* https://www.ica.org/app/uploads/2023/12/ICA_1996-09-06_code-of-ethics_EN.pdf

Ketelaar, Eric. 2001. "Tacit Narratives: The Meanings of Archives." *Archival Science*. 1: 131-141. https://doi.org/10.1007/BF02435644.

Kumbier, Alana. *Ephemeral Material: Queering the Archive.* Litwin Books, Sacramento. 2014.

MacNeil, Heather. "Archivalterity: Rethinking Original Order." *Archivaria* 66 (1): 1–24. https://archivaria.ca/index.php/archivaria/article/view/13190.

Mandarino, Grant G. 2022. "Review of 'All Shook Up': The Archival Legacy of Terry Cook." *Journal of Western Archives*. 13(1): article 10. https://doi.org/10.59620/2154-7149.1155.

Marcus, Sharon. 2013. "Queer Theory for Everyone: A Review Essay." *Signs* 31.1 (2005): 191–218. https://www.jstor.org/stable/10.1086/432743?origin=JSTOR-pdf.

Martin, Douglas. "Yvone Brill, a Pioneering Rocket Scientist, Dies at 88." *New York Times*, March 30. https://www.nytimes.com/2013/03/31/science/space/yvonne-brill-rocket-scientist-dies-at-88.html

Mason, Kären and Tanya Zanish-Belcher. 2007. "Raising the Archival Consciousness: How Women's Archives Challenge Traditional Approaches to Collecting and Use, Or, What's in a Name?" *Library Trends*. 56 (2): 344-39. https://doi.org/10.1353/lib.2008.0003.

Murari, Krishnan. 2024. "Gandhi's horoscope to Mountbatten Partition papers: Race to digitise National Archives" *The Print*, July 1.

https://theprint.in/feature/around-town/gandhis-horoscope-to-mountbatten-partition-papers-race-to-digitise-national-archives/2154359/

Neal, Jessica C. 2023. "Fleckner Revisited: Reflections on Being an Archivist, Then and Now." The American Archivist Reviews Portal. https://reviews.americanarchivist.org/2023/02/22/fleckner-revisited-reflections-on-being-an-archivist-then-and-now/

Rao, Kavitha. 2021. *Lady Doctors*. Westland.

Ridener, John. 2009. *From Polders to Postmodernism: A Concise History of Archival Theory*. Litwin Books.

Schwartz, Joan M and Terry Cook. 2002. "Archives, Records, and Power: The Making of Modern Memory." *Archival Science*. 2: 1–19. https://doi.org/10.1007/BF02435628

Skinner, Makala and Ioana Hulbert. 2022. "A*CENSUS II All Archivists Survey Report." Ithaka S+R. https://doi.org/10.18665/sr.317224.

Sullivan, Margaret. 2013. "Gender Questions Arise in Obituary of Rocket Scientist and Her Beef Stroganoff." *New York Times*, April 1. https://archive.nytimes.com/publiceditor.blogs.nytimes.com/2013/04/01/gender-questions-arise-in-obituary-of-rocket-scientist-and-her-beef-stroganoff/

The Life of Science. https://thelifeofscience.com/

Thomas, Renny. 2020. "Brahmins as scientists and science as Brahmins' calling: Caste in an Indian scientific research institute." *Public Understanding of Science*, 29 (3): 306–318. https://doi.org/10.1177/0963662520903690.

Thomas, Renny. 2021. Science and Religion in India: Beyond Disenchantment (1st ed.). Routledge.

VijayRaghavan, K., and Michael Bate. 2010. "Veronica Rodrigues (1953–

2010)." *Science* 330 (6010): 1493–1493. https://doi.org/10.1126/science.1200787.

"Visualizing Women Scientists in the APS Collections." https://diglib-legacy.amphilsoc.org/womeninscience/

Yakel, Elizabeth. 2003. "Archival representation." *Archival Science* 3: 1–25. https://doi.org/10.1007/BF02438926.

Zinn, Howard. 1977. "Secrecy, Archives, and the Public Interest." *The Midwestern Archivist.* 2 (2): 14-26. https://www.jstor.org/stable/41101382.

Endnotes

1 This paper is based on a presentation made to archivists and historians at the Fourth Workshop on Scientific Archives on 6 June 2024 in San Francisco. In that presentation, we discussed our concerns about the lack of diversity in our archive and changes we had made to our workflow in order to bring in more diverse material. Here, we build on the ideas we originally presented to also include an examination of archival theory and the process of digitizing archival material. We would like to extend our warmest thanks to Gita Chadha, Iz Paehr, S. Prashant Kumar, Samira Agnihotri and Venkat Srinivasan for commenting on this paper and providing feedback. We are especially indebted to Gita, Venkat and Prashant for their critiques that helped in formulating responses to several issues we have raised in this paper.

2 The Archives at the National Centre for Biological Sciences (NCBS), where the authors of this paper work, is a public collecting center for the history of science in contemporary India. Since 2019, the Archives has been collecting, preserving, cataloguing and making material accessible with the aim of helping its users gain insight into the practice and culture of science in India. Collections include papers related to biological sciences – as well as material related to the institutional history of the center it is based at; scientists, and other STEM fields such as mathematics, physics, space science, and ecology. The Archives also holds material at the intersection of science and society such as papers from environmental activists and protection groups, writers, artists and journalists. The Archives is funded primarily by grants from Arcadia, a charitable foundation based in the UK, and by TNQ Foundation, an Indian publishing and technology services company that provides support for the life sciences, mathematics and the arts.

3 The Obaid Siddiqi Chair was instituted at the Archives to bridge the gap between the sciences, social sciences and humanities, and foster a diverse and inclusive community of academics across disciplines as part of a broader effort in developing a program in the History and Culture of Science.

4 Material related to Veronica Rodrigues can be found in the NCBS institutional collections NI-001 - NCBS: Establishment Office Collection and NI-002 - NCBS: Office of Dean Collection as well as in the papers of two scientists who

were her colleagues at TIFR and NCBS - MS-003 KS Krishnan Papers and MS-001 - Obaid Siddiqi Papers.

5 We'd like to acknowledge the efforts of the hundred or so formal and informal members of the Archives over the years who have contributed to its ethos and work culture, conversations about gender, inclusivity, archives and the sciences. These were shaped at the archives right from its inception by Venkat Srinivasan, Janaki Nair, Indira Chowdhury, Aparna Uppaluri, Satyajit Mayor and many others – and we see our work as a continuation of these efforts.

6 In the introduction to her book foregrounding the histories of women in medicine, Kavitha Rao argues that erasing the stories of the few women who managed to persevere despite discrimination allows the argument that women have made no significant contribution to science, and that they are unfit for the job. See Rao 2021, 10.

7 This is something we have also heard anecdotally from fellow archivists in India. Sharon Marcus writes that LGBT and queer histories are even harder to trace (Marcus 2013, 201).

8 The archives policy is to align with various provisions of the Public Records Act which regulates records in the National Archives of India. It allows unclassified records older than 30 years to be made available to researchers, though the Act itself is limited to government entities. See: Archives, Ethics and the Law in India: A Guidebook for Archivists in India.

9 The Jamal Ara Papers at the Archives contain a notebook with a fictionalized account of an abusive marriage, while the Chayanika Shah Papers contain case-files mentioning incidents of assault and a legal case. Both of these folders are flagged as restricted.

10 Out of the 11 collections, four are from women scientists who had worked at leading Indian scientific institutions, the others are papers from women who worked with science as activists, artists, filmmakers, writers and journalists.

11 To demonstrate the sometimes opposing ways in how we are perceived as a science archive, we offer two examples: a male scientist who looked through our first community collection – the WPSS Papers of an environmental organization in Wayanad, Kerala, told us that "any scientist will tell you that this is not science". A prominent educator who visited the archives remarked that she

was pleasantly surprised to find material related to people-led science popularization and social movements in a "science" archive.

12 The course was an invitation to develop sociological perspectives on science, with the express intention of critically impacting on the practice of science, particularly biology. Using gender as a lens, the course attempted to present an intersectional feminist reading of the history of science, and encouraged participants to develop their own understanding of the gendered nature of science and an imagination of how to change it.

13 Grant G. Mandarino, in a review of "All Shook Up," a compilation of Terry Cook's writing on archives, writes, "Self-conscious reflexivity and a dedication to broader, more flexible contextualization are the maxims of 'postmodernism' Cook seems to have been most drawn to. Yet it remains unclear why these insights are uniquely identified as being 'postmodern,' when they could just as easily (and justifiably) be linked to intellectual trends in the humanities and social sciences dating back to the New Left and rising feminist movements of the 1960s and 1970s." See: Mandarino 2022.

14 The team at the Archives consists of archivists, conservators, administrators, a historian, an IT specialist, a facility manager and subject-matter experts. Nine out of the fourteen-member team identify as women. In our limited experience with many of the archives we have visited in India, staff are largely comprised of women, though there has been no survey that reports on archivists in India. The SAA lead survey in the USA revealed that the profession tends to have a higher proportion of women – 71% (Skinner and Hulbert 2022).

15 Cifor discusses how working toward social justice in archives requires collecting and appraising in more socially conscious ways that extend concepts of who and what is of value, "rethinking what is necessary to serve diverse constituencies ethically, and challenging dominant normative political, economic and social structures through which power, privilege and oppression are enacted. Such efforts cannot be made without a critical eye on current conditions, constraints, and possibilities that stand in the way of attaining social justice" (Cifor 2016, 9).

16 Mason and Zanish-Belcher also describe how participating in oral histories help to convince women about the significance of their role in history and the

value of their life stories, how this work can also raise the consciousness of any community about the value of its history, the collaboration between the archivist and the narrator in creating an archival record, and how in some cases this bond between the two leads to the donation of papers, when they exist. See: Mason and Zanish-Belcher 2007, 350.

17 Samira Agnihotri connected the Archives to a wildlife biology student she had been mentoring, Abhijith AV, who collected and brought the Wayanad Prakrithi Samrakshana Samithi Papers to the Archives.

18 For the obituary, see: Martin 2013. For the discussion that this generated, see: Sullivan 2013.

19 As critiques based on race do too: archivist Jessica C. Neal writes, "my positionality as a Black queer archivist took root in grad school and blossomed into the perspective that whether through theory or practice, archival work is most radical, impactful, and mission-filled when it is decolonial, collaborative, imaginative, and less concerned with output, product, and the 'howness' of the work" (Neal 2023).

20 Caswell emphasizes using archival materials "in the now" as opposed to preserving them for "some vague future" (Caswell, 2021b, 38).

21 Kumbier also cautions that in the queer archive, "[c]ontemporary Library of Congress subject headings (used to describe individual books and entire collections of papers in conventional archives) can account for some of these identifications, but they can't account for all of them in their particularity, their inventiveness, their impermanence, and their ambiguity," offering Eve Sedgwick's playful list of terms from the foreword to their 1993 work *Tendencies* as an example.

22 Boles also cites TR Schellenberg, who in 1961 wrote that the archivist "should have no compunction about rearranging series in relation to each other or single record items within them if by so doing he can make the records more intelligible and more serviceable. The test here is a very practical one of usability."

23 https://diglib-legacy.amphilsoc.org/womeninscience/. This work by David Ragnar Nelson was also presented at the Fourth Workshop on Scientific Archives at UCSF in June 2024.

24 This includes mammoth digitization projects at the National and State Archi-

ves in India. These projects require a large number of digital archivists with specialized skills, and with the archives not having enough capacity to provide training, digitization is outsourced to consultancies which employ a young labor force with little to no standardization and processes in place (Murari 2024).

25 The Archives at NCBS has launched a regrant programme for archiving the history of underrepresented groups in science in India, where the communities who are custodians or creators of the material appraise, arrange, catalogue, describe and digitise their material and integrate it to the digital archive hosted by us. In 2025, the archives will be working with about 9 selected participants to create an archive of history of science in India framed from the margins, and we hope it will inform our own approach to similar material in the archive.

11

Connections, Collaborations, Collections: Using Data Visualization to Rethink Access to Scientific Archives

David Ragnar Nelson, American Philosophical Society
Serenity Sutherland, SUNY Oswego

Abstract
Traditional modes of archival access and discovery, such as finding aids, do not always help reveal marginalized voices in collections. This paper discusses efforts to reframe archival discovery through data visualization. These approaches eschew the traditional hierarchy of finding aids and can, we argue, reveal agents otherwise hidden in collections. We focus on two efforts undertaken at the American Philosophical Society: one effort, *Visualizing Women in Science*, documents women scientists in the archives and allows users to discover information about them in an interactive network visualization; the second effort applies the lessons of *Visualizing Women in Science* to a prototype subject guide for the entirety of the collections. These projects attempt to reframe users' access to the archives away from singular collections and away from singular historical figures and instead encourage a cohesive approach to the totality of collections. While such efforts take significant amounts of labor, we argue that this reframing is a productive step in confronting and remediating archival silences.

Scientific archives provide access for researchers that illuminate the lives and scholarship of important scientific developments. Beyond this apparent factor, though, are a range of questions about access, the production of history, and whose voices are represented within an archival repository. These questions include: What is contained in a scientist's papers? Whose voice is represented? Does a scientist's collection speak to the scientist's biography or more to their social milieu? Are their papers a truthful sample of their activities, or a constructed and curated record of how they wanted to be remembered? Of course, scientific papers can be all of these things, and more. However, the assumptions with which archivists and information science professionals approach these papers may structure how we describe, curate, and promote access to them. If researchers, then, approach the archive assuming the most significant thing about a collection is the scientist whose name it bears, they will likely miss a good deal of the interstices that exist between collections. Thinking outside the traditional structures of the archive and archival description may help reveal more of these linkages.

At the American Philosophical Society (APS), we have been experimenting with efforts to read **through** the archive, viewing the archive as a cohesive collection of interrelated scientific records. These experiments draw heavily on methods from the digital humanities and seek not only to improve access to the collections, but also to make the collections available to broader audiences. In an effort to recognize the limitations imposed by archival collections and finding aids,[1] the APS has been using data visualization and digital storytelling to find commonalities between collections, filling gaps and silences in one collection with related material in another.[2]

All archival collections are incomplete.[3] Whether through the deliberate destruction of material, natural disasters, carelessness, or intentional suppression, each collection only provides a small sample of all the possible material that must have once existed. As Verne Harris has convincingly argued, the combination of deliberate destruction, loss, and archival selection means the current archive likely only represents "a sliver of a sliver of a sliver" of the documentary record that must have originally been created.[4] These gaps and silences are not even across collections or groups of people. In particular, archival silences tend to reinforce the power structures that created them, meaning that underrepresented groups such as women, ethnic and racial minorities, and contingent laborers will be more likely subject to gaps in the documentary record. The curatorial aspects of archival selection, preservation, and increasingly, choices about digitization allude to the process of production that goes into "making" an archive and the power that comes with including voices in archival records.[5] Scientific records can face additional problems in this regard, as silences can come from the scientists themselves—many scientists, both historically and today, have not understood the value of their records and therefore have made no efforts to preserve them.[6]

While we can never recover material that was not preserved, it may be possible to reframe how we think about presenting archival material to create a fuller picture of the holdings. Can we juxtapose collections to complement the missing material in one collection with what is present in another? That is, can we use one collection to respond to a silence in another? At the American Philosophical Society, we have been experimenting with modes of visual storytelling to 1. Try to confront gaps and silences in our collections, and 2. Make our collections more accessible to a wider range of audiences. This paper will discuss two related efforts. The first, *Visualizing Women in Science*, unearths the biographies and professional networks of over two hundred women scientists documented in the APS collections. The second is a preliminary description of an attempt to apply the lessons and successes of *Visualizing Women in Science* to all of the history of science collections at the APS in a revamped subject guide for the history of science.

About the American Philosophical Society

The American Philosophical Society (APS) was founded in 1743 by Benjamin Franklin with the purpose of "promoting useful knowledge."[7] As the oldest learned society in the United States, the APS today is a membership organization with a body of approximately one thousand elected members, who represent the natural sciences, social sciences, humanities, and public life. While the APS is an expansive institution today, Franklin's original idea for the APS was focused squarely on natural philosophy, encompassing fields such as chemistry, botany, medicine, and mathematics. Because of this, the institution's history attests to scientific activity in the United States over its entire history, containing rich records in meteorology, geology, engineering, the life sciences, and the physical sciences. Since the beginning, the APS has maintained a library on these subjects, with an increasing interest in collecting the papers of leading scientists beginning in the mid-twentieth century.[8] Today, the library has over 14 million pages of manuscript material in addition to a large printed book collection. The APS's library has three main collecting areas: early American history, Indigenous languages and cultures of the Americas, and the history of science.

The history of science comprises the single largest collecting area for manuscripts, making up just over 70% of the holdings by volume. The collections span the entirety of the institution's history—or as some staff members have termed it, from "Newton to NASA"—but the manuscript holdings are particularly concentrated in the twentieth century. The collections are particularly strong in physics and the life sciences, with the papers of five Nobel laureates in the biological and medical sciences, including pathologist Peyton Rous, geneticist Barbara McClintock, and former APS president Baruch Blumberg, and two Nobel laureates in physics.

In recognition of the unique concentration of scientific papers at the APS, the APS founded a Center for the History of Science in July 2024.[9] The projects described in this paper will inform the future development of both the center and the scientific collections at the APS.

Yet the history of science collections present researchers with unique, persistent challenges. While 18th century collections are cataloged at the item level, 20th century collections are only processed to the file level.[10] This means researchers often have difficulty finding specific pieces of information. This is compounded by the fact that history of science collections tend to be quite large. Researchers may have to sift through boxes and boxes of correspondence to find all information relevant to their research question.

Additionally, the level of processing these collections have received can vary widely, even among collections with similar provenance. A substantial subset was processed flatly and has not been broken out into different series. Some have only paper inventories, or no inventories at all. Given the current backlog and the current collection priorities, it is unlikely that the condition of these collections will be improved anytime soon. It is almost certain that valuable information and narratives are hidden in plain sight in neglected collections.

Furthermore, due to the rich concentration of related materials at the APS, researchers often benefit from consulting multiple collections. For instance, the APS holds the papers of several researchers affiliated with the Rockefeller Institute, including Florence Sabin and Peyton Rous. Staff at the APS have identified several instances in which key details about one researcher are only divulged in the papers of one of their colleagues.[11] If researchers are not encouraged and even empowered to consult these collections in tandem, they will miss critical details and connections. In particular, traces of marginal figures are more likely to be broken up across collections, leaving traces here and there. As collecting priorities have emphasized collecting the papers of students who worked with the scientists whose papers the APS already holds, it is almost certain that this pattern will continue.

Visualizing Women in Science

The project *Visualizing Women in Science* marks a first attempt at rethinking how the APS presents collections.[12] *Visualizing Women in Science* was a multi-year collaborative effort to more fully document the presence of women scientists in the collections. Funds provided via the APS Library & Museum digital humanities fellowship program allowed a historian, Serenity Sutherland, to comb through the archives to search for women scientists mentioned in more prominent women scientist's collections.[13] Ultimately the goal of this detailed archival work was to recover the histories of other women of science who lack complete records and whose scientific contributions are very little known.

Another key goal of reading through the archive and noting any mention of women scientists was to challenge the "heroine" narrative that often accompanies famous women scientists, or what Julie des Jardins has called the "Madame Curie Complex."[14] This refers to the tendency to focus on a small number of "superstar" women scientists at the expense of others in the field. The visualization builds on the work of historians of women in science who sought women's records in the archives.[15] The project attempts to push further into archival discovery to perform a close investigation of

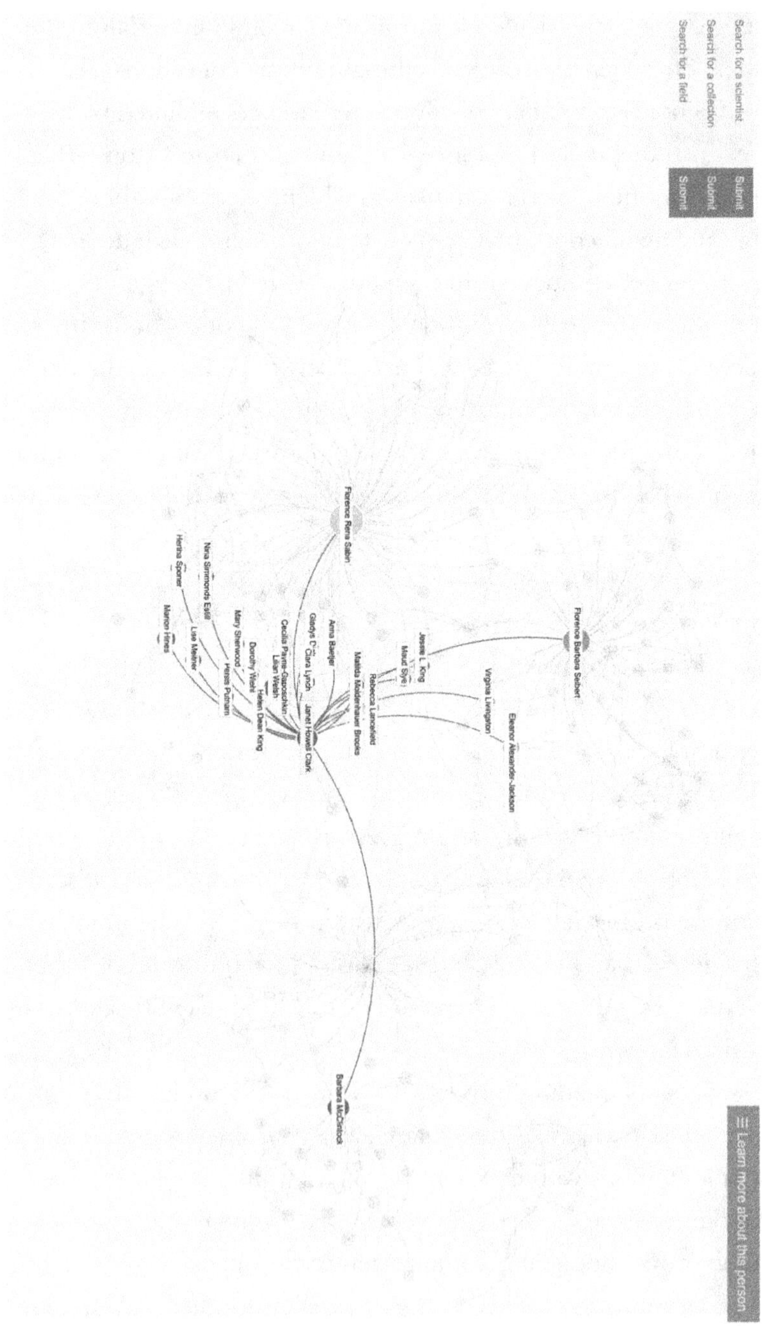

Figure 1: The network visualization from *Visualizing Women in Science*. Users can click on nodes to reveal connections in the network

the larger network of actors around more famous women scientists. Doing so would reveal a richer picture of the paths that were available to women scientists. We also hoped that the project would uncover and recover narratives and biographies of marginal women scientists. By presenting the work as a network visualization, we would avoid a hierarchical organization of the scientists featured.[16]

The project is focused largely on six APS collections. Five of these—the papers of anatomist Florence Sabin, the papers of biochemist Florence Seibert, the papers of geneticist Barbara McClintock, the papers of physicist Rose Mooney-Slater, and the papers of biophysicist and biochemist Mildred Cohn—are the papers of women scientists. The last collection, the Jackson Laboratory Oral History Collection, was selected due to the high concentration of women geneticists who worked in that lab. These papers were selected because, at the time, they represented all the collections of women scientists available in a processed form. One additional collection, the papers of biochemist Jenny Glusker, became available during the course of the project, but this collection did not produce enough meaningful material to incorporate. Future collections, such as the papers of embryologist Beatrice Mintz Papers, may be incorporated as resources allow. In addition, we did targeted searches in other collections to find more connections. The papers Britton Chance, who worked closely with Mildred Cohn but preserved many more records, were particularly useful in fleshing out Cohn's network.

These five papers at the project's core represented the only available papers of women scientists at the project's start. The fact that such a small percentage of the holdings represented women's contributions to science indicates both the urgency of the project and also the need to address women's contributions explicitly in future collecting. Of these five papers, two came to the APS by happenstance: the Florence Sabin papers were part of a larger gift from the Rockefeller Institute,[17] and the Rose Mooney Slater papers were donated following the papers of her husband, John Clarke Slater. This means that only three of the collections in the project represented explicit attempts to gather the papers of women scientists. As such, the project provided the Society with an opportunity to reaffirm its commitment to providing a good home going forward for records about women in science.[18] Two related efforts coincided with the project, further emphasizing this point: *Visualizing Women in Science* was launched in 2023 to coincide with an exhibition at the APS museum on women in science, entitled *Pursuit and Persistence: 300 Years of Women in Science*,[19] and also coincided with an effort to collect oral histories from living women APS members in the sciences for the collections.[20]

The method used was the following: the historian went folder by folder through the correspondence of the five core collections of women scientists, paying particular attention to folders with women's names.[21] Every time she was able to document a plausible connection between two women scientists, she recorded this in the dataset. The types of connections she looked for included the following: correspondence; mentorship relationships; working in a lab together; coauthoring a paper; serving on a committee together; writing and receiving letters of recommendation about a person; and asking for recommendations. These relationships were chosen to provide the network with some degree of cohesion and to ensure that the network modeled like relationships.[22] After this period, staff members at the APS performed targeted searches in other collections. It was at this point that the Jackson Laboratory Oral History Collection was incorporated after rich interconnections with other collections were discovered. The resulting dataset had 494 connections and 259 individuals. Since the project was always intended as a piece of public-facing digital scholarship, eight men were included on account of their fame or significance in the APS collections.

After the initial dataset was constructed, we authored biographies for all individuals in the dataset. These biographies were written using a variety of sources, including the archival holdings, but also web sources such as obituaries, digitized university graduation records, and digitized newspaper articles. In many cases, the biographies are incredibly sparse as little information was available on the scientist. Each scientist's biography exists as a node within the visualization. A line between two nodes, called an edge, indicates that the APS holdings document some sort of relationship between the two nodes. The types of connections we documented were oftentimes not evident from the finding aid. The name of the person who ended up in the network is not always the same as the name that appears on the folder. In some particular cases, we consulted multiple folders, or even multiple collections, to form a cohesive picture of the person. For this reason, the resulting network is necessarily fairly small and only produces limited meaningful statistical results.[23] Nonetheless, the visualization provides a means for thinking about the figures documented in these collections in a novel way.

For one, the connections in the network only loosely correlate with collections. Nodes are colored algorithmically using Clauset-Newman-Moore greedy modularity maximization.[24] This method finds groups of nodes that have a higher number of connections between each other than they do with any other part of the network. While these communities generally correspond with collections, they occasionally reveal other trends. For instance, connections originally derived from Florence Sabin's collections ended up in one of three communities. One is a comparatively loose network made up of women scientists who only occasionally corresponded with Sabin. The other is a dense portion of the network centered on Sabin's work with the Naples Table Association and the Emmy Noether Memorial Fund. This dense section represented Sabin's most intense attempts at promoting women's success in science and were supplemented only occasionally with materials from other collections. A third part of her network was grouped with the connections found in the Florence Seibert papers. Seibert and Sabin worked closely together on tuberculosis research, and it is therefore not surprising that these connections were assigned to the smaller, more thematically cohesive Seibert section of the network. In a significantly larger network, it seems reasonable to assume these communities would reveal interesting trends across the collections.

The network found a few surprise figures hiding in the collections. For example, the project uncovered Mexican physician Consuelo Vadillo Gutierrez from the Yucatán Peninsula who the APS's collections describe as "the first woman doctor" in the Yucatán.[25] She studied briefly in the

United States at Johns Hopkins University and at the Women's Medical College of Pennsylvania. The Florence Sabin Papers contain records of her course of study in the United States, but this material is divided between two folders: one in the folder for Kathryn McHale, the general director of the American Academy of University Women, and one in the folder for Heloise Brainerd, a representative of the Pan American Union. Vadillo is completely absent from the finding aid and it was only the attempt to think through the collection relationally that revealed her presence. Though the correspondence does not contain enough information to tell a full biography of Vadillo, it does provide a crucial lead for telling more of her story.

Another example is the case of Elizabeth Eshelman Miller. Miller wrote her dissertation in the 1950s at the University of Pennsylvania with tuberculosis researcher Florence Seibert. While working with Seibert, she was part of a research team that included several other women, all of whom are documented in the network. For many of these women, little information exists about what kind of careers they may have had after their time in Seibert's lab. But Miller, after a brief stint at two other hospitals, returned to the University of Pennsylvania later, where she became a colleague of Mildred Cohn's. This means Miller is also documented in Cohn's papers, as well as the papers of Britton Chance. With this information, one could begin to tell a fairly complete story about Miller's career, and perhaps also say something interesting about the evolution of the institution she worked at. In fact, we were somewhat surprised to see an overlap between Cohn and Seibert. Although both worked in biochemistry, and although both worked at Penn, on the surface, their research seemed quite different. Seibert worked on tuberculosis research, identifying and isolating tuberculin and paving the way for the development of the TB test, and later worked on the possibility of a bacterial origin for cancer. Cohn, on the other hand, was a pioneer in nuclear magnetic resonance and studied enzyme reactions. Seibert seems to fit more comfortably in the lineage of early twentieth century pathologists, like those that populate Florence Sabin's network, while Cohn belongs to a later paradigm. Nonetheless, using these two collections in tandem and pointing out the moments where they do overlap may provide a fruitful way for rethinking these histories.

Moreover, this project demonstrated that our collections are in meaningful dialogue with each other. With the exception of Florence Seibert and Florence Sabin, none of the women scientists whose collections we used worked terribly closely with the others. The closest connections they formed are mostly found within their own collections. However, we still did find meaningful connections across the archive. Additionally, we also saw how more complete collections could help augment collections that had not been as well preserved. Britton Chance's papers contained many illuminating details about women in Mildred Cohn's orbit. And for Barbara McClintock, who destroyed much of her correspondence from before 1970, we found a wealth of information about her network across the archive's collections in genetics, including in the papers of Curt Stern, Ernst Caspari, Theodosius Dobzhansky, and Salvador Luria.

In at least one case, this process also drew attention to neglected collections. For instance, cell biologist Margaret Reed Lewis emerged as a key figure in the networks of Florence Sabin and Florence Seibert. As we started to dig into other papers she might be in, we discovered that the APS holds the papers of her husband, Warren Harmon Lewis. This collection only has a paper inventory and is therefore essentially undiscoverable to users. As there is only a shell finding aid, any people documented there will not appear when users use the APS's search interface to query the collections. As we began to dig into the Lewis papers, we discovered not only that many other women in the network were there, but also that many of the letters were either from or addressed to Margaret Reed Lewis rather than her husband. While our project did not have the resources to create an electronic inventory for the papers, or to reprocess the papers to uncover the extent to which Margaret Reed Lewis was represented there, it sheds light on their existence and will hopefully help guide researchers towards them in the future.

Towards a New Subject Guide for the History of Science

Visualizing Women in Science focused on a curated selection of collections.

How might this work be applied to all of the collections in a repository? What risks could that entail? A crucial aspect of this approach was careful and deliberate attention to the collections. The slow, painstaking research the historian performed provided a method for thinking about the diversity represented in the collections. While the APS does not have the resources to use this approach on all collections, we want to argue that in the cases of groups who are more likely to be marginalized in the historical record, it is worth devoting these resources to uncovering their presence. Thinking of ways to present collections outside of traditional finding aids does not replace the slow recovery work to find these figures, but it may provide a framework for rethinking how we help researchers use collections and frame their own interactions with the holdings.

For the APS, this work holds some urgency. As part of the founding of a Center for the History of Science, the APS wants to modernize and update its subject guides for the history of science, and already we are thinking about how we can incorporate the lessons of *Visualizing Women in Science* into this work.

Currently, the subject guides are badly out of date. Each scientific field has its own guide, which prevents researchers from seeing potential overlaps between different parts of the collection. Some are drawn from the database of finding aids and can be updated, while others are digital versions of decades-old printed guides with no easy means to change. Additionally, in the absence of dedicated staff for the history of science, it was previously unclear who had the responsibility to update the guides.

Rather than fixing these old guides, the APS has decided to rethink the subject guides for the history of science from the ground up. The plan is to develop a unified subject guide for all of the history of science, with features to allow users to filter collections based on particular disciplines or research concerns. There are many open questions about this subject guide—not least of all, which collections to include. For instance, many collections in the history of anthropology are clearly in conversation with the collections in the history of biology and medicine. However, these anthropological collections are largely already covered by our Indigenous Subject Guide. Do we need to have them in both places? Do we include collections like the

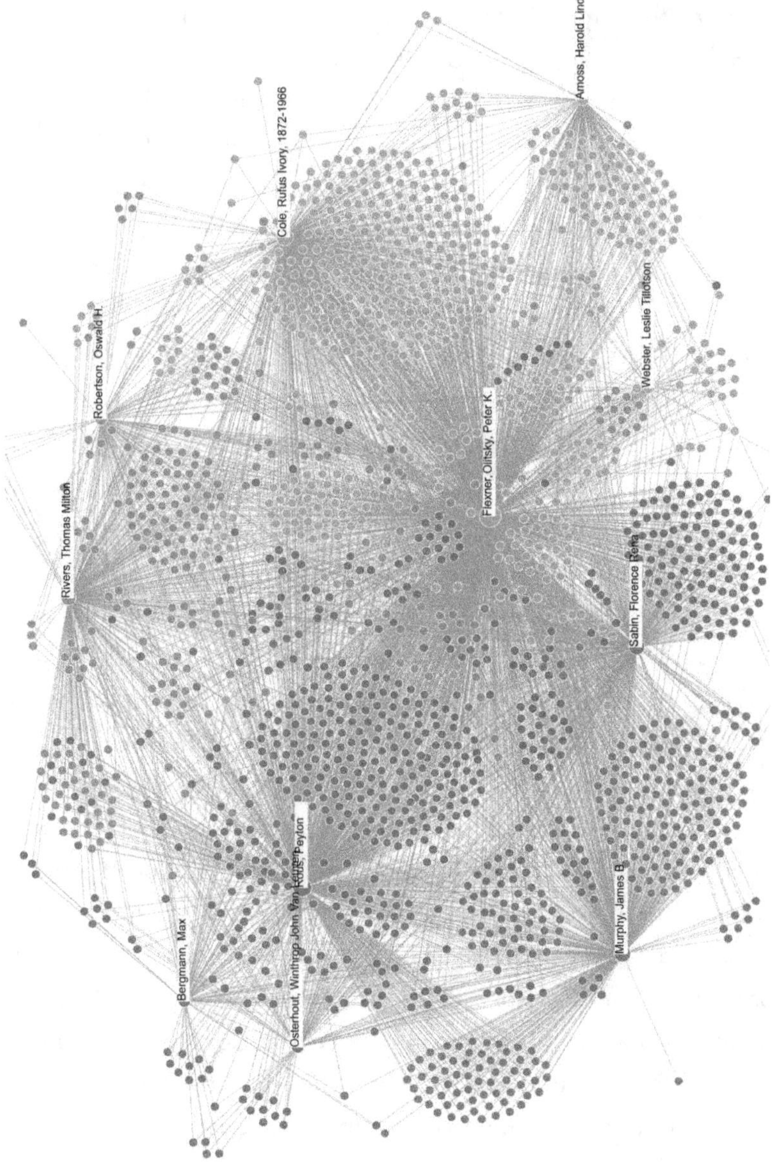

Figure 2: A prototype of what a visual subject guide could look like. Labeled nodes indicate scientists whose papers the APS holds, while other nodes are individuals named in the finding aids. A line between two nodes indicates correspondence exists between the two nodes

Benjamin Franklin Papers? Franklin was a key scientist of his era, but only a portion of the papers reflect his interest in science. How do we include unprocessed or underprocessed collections?

For now, we are restricting our initial forays into twentieth century biology, medicine, and physics. These represent main collection strengths of the APS and are also the collections most in need of attention. The main question at this stage is how we develop a sustainable workflow for developing a resource of this size. For *Visualizing Women in Science*, we had the luxury of reading every letter, finding details and names that were not captured in the finding aid. We will not have that luxury for this project if we want to finish it. We will need to develop a method to fix the most egregious processing errors, while still mainly working with readily available data.[26]

One strategy we hope to employ is augmenting the data available at the APS with data available as Linked Open Data, such as Wikidata.[27] Using these databases, we can assign disciplines and other useful information such as academic affiliation that may help researchers identify folders of interest to them. This will build upon filtering functions that we pioneered in *Visualizing Women in Science*, but will not require archivists to manually populate the information. Providing our data as Linked Open Data will also allow other archives to contribute their data, paving the way for cross-archival resources.

An initial internal prototype using twelve collections of researchers affiliated with the Rockefeller Institute has already produced some interesting results. These twelve collections, which include the papers of Florence Sabin used in *Visualizing Women in Science*, mostly came to the APS in the 1960s directly from the Rockefeller Institute. Two of them, the papers of Sabin and the papers of Nobel laureate Peyton Rous, clearly belong to the treasures of the APS and have received much attention from researchers. Another collection, the papers of Simon Flexner, is among our most used collections and at around 115 linear feet is one of our largest collections. While all of the collections have electronic inventories, many are badly in need of updating and contain many errors and inaccuracies. The extent to which these can be fully addressed in the framework of this project is an open question, but we hope that finding similarities between the collections will point to the areas most in need of attention.

The initial proof-of-concept has already shown us the value of this type of project, especially as a means of identifying people in the collection. We were able to fix several outdated records and find common correspondence between several records. In doing so, we are starting to have a sense of which of these collections are the most similar, and which might be best used in tandem. For instance, our efforts have shown that the Flexner collection, which is significantly larger than any other collection included, is a necessary auxiliary source for all of these collections. It also evaluated the only woman scientist, Florence Sabin, in a separate algorithmic community, reinforcing there may be something different about her collection. But much work is still needed to turn this into a usable resource, and to make it into something we can use across all the collections. With the establishment of the Center for the History of Science and the allocation of two full-time staff members to assist in this and related efforts, the APS hopes to develop a sustainable model for pursuing this type of work.

Bibliography

Abir-Am, Pnina G. and Dorinda Outram. *Uneasy Careers and Intimate Lives: Women in Science, 1789-1979*. New Brunswick, NJ: Rutgers University Press, 1987.

Ahnert, Ruth, Sebastian E. Ahnert, Catherine Nicole Coleman, and Scott B Weingart. "Visual Networks." In *The Network Turn: Changing Perspectives in the Humanities*. Cambridge: Cambridge University Press: 2020. https://www.cambridge.org/core/elements/network-turn/CC38F2EA9F51A6D1AFCB7E005218BBE5.

The American Philosophical Society Year Book 1943. Philadelphia: American Philosophical Society, 1944.

Anderson, Bethany G., and Kristen Allen Wilson. "Remembering Women Scientists: The Case for Proactively Documenting Women in Science." *Collections: A Journal for Museum and Archives Professionals* 18, no. 4 (2022): 453-78.

"APS Establishes Center for the History of Science." American Philosophical Society. Last modified November 5, 2024. https://www.amphilsoc.org/blog/aps-establishes-center-history-science.

Berry, Dorothy. "Take Me into the Library and Show Me Myself: Toward Authentic Accessibility in Digital Libraries." *Transaction of the American Philosophical Society* 110, no. 3 (2022): 111-126.

Berry, Dorothy. "What's in a Finding Aid and What's Not." *The Hows and Whys of Finding Aids*. Last modified July 6, 2020. https://scalar.fas.harvard.edu/finding-your-way-through-finding-aids/whats-in-a-finding-aid-and-whats-not.

Boyd, Jane and Magdalena Hoot. "Pursuit & Persistence: 300 Years of Women in Science (Online Exhibition)." American Philosophical Society, Last modified January 11, 2024. https://www.amphilsoc.org/museum/exhibitions/pursuit-persistence.

Bystydzienski, Jill M., and Sharon R. Bird, eds. *Removing Barriers: Women in Academic Science, Technology, Engineering, and Mathematics*. Bloomington: Indiana University Press, 2006.

Cairo, Alberto. *The Truthful Art: Data, Charts, and Maps for Communication*. New Riders/Pearson, 2016.

Carter, Rodney G. S. "Of Things Said and Unsaid: Power, Archival Silences, and Power in Silence." *Archivaria* 61 (2006): 215-33. https://archivaria.ca/index.php/archivaria/article/view/12541.

Catlett, J. Stephen. *A New Guide to the Collections in the Library of the American Philosophical Society*. Philadelphia: American Philosophical Society, 1987.

Clauset, Aaron, M. E. J. Newman, and Cristopher Moore, "Finding community structure in very large networks," *Physical Review E* 70 (December, 2004): 066111-1-066111-6, https://doi.org/10.1103/PhysRevE.70.066111.

des Jardins, Julie. *The Madame Curie Complex: The Hidden History of Women in Science*. New York: Feminist Press at the City University of New York, 2010.

Doel, Anna. "Women in Science—Oral Histories from APS Members." American Philosophical Society. Last modified August 26, 2024. https://www.amphilsoc.org/museum/exhibitions/women-science-

-oral-histories-aps-members.

Drucker, Johanna. "Graphical Approaches to the Digital Humanities." In *A New Companion to the Digital Humanities*, edited by Susan Schreibman, Ray Siemans, and John Unsworth. Wiley-Blackwell: 2016.

Fisher, Lillian Estelle. "The Influence of the Present Mexican Revolution upon the Status of Mexican Women." *The Hispanic American Historical Review* (1942): 221–28.

Franklin, Benjamin. "A Proposal for Promoting Useful Knowledge, 14 May 1743," In *The Papers of Benjamin Franklin*, edited by Leonard W. Labaree, 378-383. Volume 2 of *The Papers of Benjamin Franklin*. New Haven: Yale University Press, 1961.

Friis, Herman R. "Graphic Approach Offers a New Approach to the Archivist's Finding Aid Program." *The American Archivist* 15, no. 2 (1952): 157-58. https://doi.org/10.17723/aarc.15.2.jj16g66436256pt8.

Glass, Bentley. *A Guide to the Genetics Collections of the American Philosophical Society*. Philadelphia: American Philosophical Society Library, 1988.

Greene, Mark A., and Dennis E. Meissner, "More Product, Less Process: Revamping Traditional Archival Processing." *American Archivist* 68 (2005): 208-263. https://doi.org/10.17723%2Faarc.68.2.c741823776k65863.

Harding, Sandra. *Whose Science? Whose Knowledge?: Thinking from Women's Lives*. Ithaca, NY: Cornell University Press, 1991.

Harris, Verne. "The Archival Sliver: Power, Memory, and Archives in South Africa." *Archival Science* 2 (2002): 63–86. https://doi.org/10.1007/978-94-010-0570-8_9.

Hawkins, Ashleigh. "Archives, linked data and the digital humanities: increasing access to digitised and born-digital archives via the semantic web." *Archival Science* 22 (2022): 319-344.

Jaillant, Lise. "More Data, Less Process: A User-Centered Approach to Email and Born-Digital Archives." The American Archivist 85, no. 2 (2022): 533-555, https://doi.org/10.17723/2327-9702-85.2.533.

Jenkins, Nicholas, Elijah Meeks, and Scott Murray. "Kindred Britain." Stanford University. Last modified 2011. https://kindred.stanford.edu/.

Jones, Huw, and Yasmin Faghihi, "Manuscript Catalogues as Data for Research: From Provenance to Data Decolonisation." *Digital Humanities Quarterly* 18, no. 3 (2024). https://www.digitalhumanities.org/dhq/vol/18/3/000751/000751.html.

Kay, Lily E. *Molecules, Cells, and Life: An Annotated Bibliography of Manuscript Sources on Physiology, Biochemistry, and Biophysics, 1900-1960, in the Library of the American Philosophical Society*. Philadelphia: American Philosophical Society Library, 1989.

Keller, Evelyn Fox. *Reflection on Gender and Science*. New Haven, CT: Yale University Press, 1985.

Keller, Evelyn Fox. *Secrets of Life, Secrets of Death: Essays on Language, Gender, and Science*. New York: Routledge, 1992.

Koeser, Rebecca Sutton, Brian Croxall, Alice Hickcox et al. "Belfast Group Poetry | Networks." Emory Center for Digital Scholarship. Accessed January 9, 2024. https://belfastgroup.ecds.emory.edu/.

Kuhn, Thomas, John L. Heilbron, Paul Forman, and Lini Allen. *Sources for History of Quantum Physics: An Inventory and Report*. Philadelphia: American Philosophical Society, 1967.

Lima, Manuel. *Visual Complexity: Mapping Patterns of Information*. New York: Princeton Architectural Press, 2011.

Lynch, Tom J. "Social Networks and Archival Context Project: A Case Study of Emerging Cyberinfrastructure." *Digital Humanities Quarterly* 8, no. 3 (2014). https://digitalhumanities.org/dhq/vol/8/3/000184/000184.html.

Merchant, Carolyn. *The Death of Nature: Women, Ecology and the Scientific Revolution*. New York: Harper One, 1990.

Moss Michael, and David Thomas, eds. *Archival Silences: Missing, Lost and Uncreated Archives*. New York: Routledge, 2021.

Nurmikko-Fuller, Terhi. *Linked Data for Digital Humanities*. New York: Routledge, 2024.

Owens, Trevor, and Jesse Johnston. "Archivists as Peers in Digital Public History." In *Handbook of Digital Public History*, edited by Serge Noiret and Mark Tebeau, 149-164. Berlin and Boston: De Gruyter Oldenbourg, 2022.

Painter, Deryc T., Bryan C. Daniels, and Jürgen Jost. "Network Analysis for the Digital Humanities: Principles, Problems, Extensions." *Isis* 110, No. 3 (September 2019): 538-554.

Pattuelli, M. Cristina, Matt Miller et al. "Linked Jazz." Semantic Lab at Pratt. Last modified March 11, 2020. https://linkedjazz.org.

"Pursuit & Persistence: 300 Years of Women in Science Opens at APS." American Philosophical Society. Last modified March 30, 2023. https://www.amphilsoc.org/blog/pursuit-persistence-300-years--women-science-opens-aps.

Rossiter, Margaret W. "The More You Look, the More You Find: Archives of Recent American Women Scientists." *Centaurus* 54, no. 4 (2012): 288-291.

Rossiter, Margaret W. *Women Scientists in America: Struggles and Strategies to 1940*. Baltimore: Johns Hopkins University Press, 1982.

Schofield, Tom, David Kirk, Telmo Amaral et al. "Archival Liveness: Designing with Collections Before and During Cataloguing and Digitization," *Digital Humanities Quarterly* 9, no. 3 (2015). https://digitalhumanities.org/dhq/vol/9/3/000227/000227.html.

Sorensen, Amanda H., Camila Escobar-Vredevoogd, Travis L. Wagner and Katrina Fenlon. "Recovering and Reusing Historical Data for Science: Retrospective Curation Practices Across Disciplines." In *Information for a Better World: Normality, Virtuality, Physicality, Inclusivity. iConference 2023. Lecture Notes in Computer Science*, edited by Isaac Sserwanga, Anne Goulding, Heather Moulaison-Sandy, et al. Berlin: Springer, 2023.

Sutherland, Serenity, David Ragnar Nelson, and Bayard L. Miller. "Visualizing Women in Science." American Philosophical Society, Last modified July 24, 2024. https://diglib.amphilsoc.org/womeninscience.

Tolley, Kim. *The Science Education of American Girls: A Historical Perspective*. New York: Routledge, 2003.

Trouillot, Michel-Rolph. *Silencing the Past: Power and the Production of History*. Boston: Beacon Press, 1995.

von Oertzen, Christine. *Science, Gender, and Internationalism: Women's Academic Networks, 1917-1955*. New York: Palgrave-Macmillan Press, 2012.

Wiedeman, Gregory. "The Historical Hazards of Finding Aids." *The American Archivist* 82, no. 2 (2019): 381–420. https://doi.org/10.17723/aarc-82-02-20.

Wise, M. Norton. "Making Visible." *Isis* 97, no. 1 (2006): 75-82. https://doi.org/10.1086/501101.

Zanish-Belcher, Tanya. "Documenting the Sometimes Invisible: Working With Women

Scientists." *Humanities Collections* 1, no. 4 (2001): 3-17. https://doi.org/10.1300/J139v01n04_02.

Zer-Aviv, Mushon. "If everything is a network, nothing is a network." *Visualizing Information for Advocacy*. Last modified January 8, 2016. https://visualisingadvocacy.org/node/739.html.

Endnotes

1 As collection access increasingly becomes digital, the utility of traditional finding aids has been questioned. For instance, Gregory Wiedeman has noted that archivists often conflate finding aids with access and that finding aids can inhibit access for inexperienced researchers. See Gregory Wiedeman, "The Historical Hazards of Finding Aids," *The American Archivist* 82, no. 2 (2019): 381–420, https://doi.org/10.17723/aarc-82-02-20. Likewise, Dorothy Berry has noted that finding aids often focus on what she terms the "ofness" of an item—who wrote it and when—, while some researchers would be better served by the "aboutness" of the item—what subjects it addresses; for Berry, this obfuscation can contribute to archival silences. See Dorothy Berry, "What's in a Finding Aid and What's Not," *The Hows and Whys of Finding Aids*, last modified July 6, 2020, https://scalar.fas.harvard.edu/finding-your-way-through-finding-aids/whats-in-a-finding-aid-and-whats-not.

2 The idea of using visualization as a supplement to finding aids occurred as early as 1952, when Herman R. Friis suggested using graphic modes of presentation for displaying archival records. See Herman R. Friis, "Graphic Approach Offers a New Approach to the Archivist's Finding Aid Program," *The American Archivist* 15, no. 2 (1952): 157-58, https://doi.org/10.17723/aarc.15.2.jj16g66436256pt8. Similarly, within the history of science, visualizations have been noted as useful for the work of scientists themselves as they try to communicate to the larger public the results of their work via maps, optical instrumentation, graphs and charts, and so on; see M. Norton Wise, "Making Visible" in *Isis* 97, no. 1 (2006): 75-82, https://doi.org/10.1086/501101. Several more recent projects make use of networks for displaying archival collections. See, for instance, Rebecca Sutton Koeser, Brian Croxall, Alice Hickcox et al, "Belfast Group Poetry | Networks," Emory Center for Digital Scholarship, accessed January 9, 2024, https://belfastgroup.ecds.emory.edu/; Tom J. Lynch, "Social Networks and Archival Context Project: A Case Study of Emerging Cyberinfrastructure," *Digital Humanities Quarterly* 8, no. 3 (2014), https://digitalhumanities.org/dhq/vol/8/3/000184/000184.html; and Tom Schofield, David Kirk, Telmo Amaral et al, "Archival Liveness: Designing with Collections

Before and During Cataloguing and Digitization," *Digital Humanities Quarterly* 9, no. 3 (2015), https://digitalhumanities.org/dhq/vol/9/3/000227/000227.html. Within the digital humanities, much good work on the power of visualizations have emerged as have actual network visualizations themselves, such as Linked Jazz and Kindred Britain, to name a few. See M. Cristina Pattuelli, Matt Miller et al, "Linked Jazz," Semantic Lab at Pratt, last modified March 11, 2020, https://linkedjazz.org and Nicholas Jenkins, Elijah Meeks, and Scott Murray, "Kindred Britain," Stanford University, last modified 2011, https://kindred.stanford.edu/.

3 For an overview on the problem of gaps and silences in archival records with particular attention to power, see Rodney G. S. Carter, "Of Things Said and Unsaid: Power, Archival Silences, and Power in Silence," *Archivaria* 61 (2006): 215-33, https://archivaria.ca/index.php/archivaria/article/view/12541. For a larger overview of global issues related to gaps and silences see Michael Moss and David Thomas, eds., *Archival Silences: Missing, Lost and Uncreated Archives* (New York: Routledge, 2021). On archival silence as it pertains to women in science in particular, see Tanya Zanish-Belcher, "Documenting the Sometimes Invisible: Working With Women Scientists," *Humanities Collections* 1, no. 4 (2001): 3-17, https://doi.org/10.1300/J139v01n04_02.

4 Verne Harris, "The Archival Sliver: Power, Memory, and Archives in South Africa." Archival Science 2 (2002): 63–86; here, 65, https://doi.org/10.1007/978-94-010-0570-8_9.

5 Michel-Rolph Trouillot, *Silencing the Past: Power and the Production of History* (Boston: Beacon Press, 1995).

6 The preservation of scientific data and correspondence has long been a problem, and the APS has undertaken multiple efforts to combat it. For instance, in the 1960s, the Society convened a committee in partnership with the American Physical Society to document and preserve sources in quantum physics. See Thomas Kuhn, John L. Heilbron, Paul Forman, and Lini Allen, *Sources for History of Quantum Physics: An Inventory and Report* (Philadelphia: American Philosophical Society, 1967), v-ix. A similar committee was convened for sources in genetics in 1977. See Bentley Glass, *A Guide to the Genetics Collections of the American Philosophical Society* (Philadelphia: American Philosophical Society

Library, 1988), x-xii. More recently, there has been a wider acknowledgment of the fragility of scientific data and the need to preserve it going forward. For an overview of current challenges, see Amanda H. Sorensen, Camila Escobar-Vredevoogd, Travis L. Wagner and Katrina Fenlon, "Recovering and Reusing Historical Data for Science: Retrospective Curation Practices Across Disciplines", in, *Information for a Better World: Normality, Virtuality, Physicality, Inclusivity. iConference 2023. Lecture Notes in Computer Science*, eds. Isaac Sserwanga et al (Berlin: Springer, 2023).

7 Benjamin Franklin, "A Proposal for Promoting Useful Knowledge, 14 May 1743," in *The Papers of Benjamin Franklin*,ed. Leonard W. Labaree (New Haven: Yale University Press, 1961), 2:380.

8 Modern collecting in the history of science began in earnest following a 1943 report of the committee on the library. See *The American Philosophical Society Year Book 1943* (Philadelphia: American Philosophical Society, 1944), 60-4.

9 "APS Establishes Center for the History of Science," American Philosophical Society, last modified November 5, 2024, https://www.amphilsoc.org/blog/aps-establishes-center-history-science.

10 With the great influence of minimal processing on recent archival practice, processing to the folder level has become a widespread practice. See Mark A. Greene and Dennis E. Meissner, "More Product, Less Process: Revamping Traditional Archival Processing," *American Archivist* 68 (2005): 208-263, https://doi.org/10.17723%2Faarc.68.2.c741823776k65863. Increasingly, there are calls to apply the tenets of "More Product less Process" to digital collections as well. See Lise Jaillant, "More Data, Less Process: A User-Centered Approach to Email and Born-Digital Archives," *The American Archivist* 85, no. 2 (2022): 533-555, https://doi.org/10.17723/2327-9702-85.2.533. However, as access to archival materials is increasingly digitally mediated, concern has been raised on whether minimal processing can adequately allow for discovery of marginalized groups. See Dorothy Berry, "Take Me into the Library and Show Me Myself: Toward Authentic Accessibility in Digital Libraries," *Transaction of the American Philosophical Society* 110, no. 3 (2022): 111-126.

11 The rich interconnections in the life sciences collections of the APS has been apparent since at least the late 1980s, when Lily Kay included a section of

scientists documented in other scientists' collections in her guide to the APS's life science collections; however, Kay does not include any women scientists in this section. See Lily E. Kay, *Molecules, Cells, and Life: An Annotated Bibliography of Manuscript Sources on Physiology, Biochemistry, and Biophysics, 1900-1960, in the Library of the American Philosophical Society* (Philadelphia: American Philosophical Society Library, 1989), 64-79.

12 Serenity Sutherland, David Ragnar Nelson, and Bayard L. Miller, "Visualizing Women in Science," American Philosophical Society, last modified July 24, 2024, https://diglib.amphilsoc.org/womeninscience.

13 The close collaboration of archivists and historians is emerging as a key element of improving public history projects. See Trevor Owens and Jesse Johnston, "Archivists as Peers in Digital Public History," in *Handbook of Digital Public History*, eds. Serge Noiret and Mark Tebeau, 149-164 (Berlin and Boston: De Gruyter Oldenbourg, 2022).

14 Des Jardin's notion of the Madame Curie Complex derives at least in part from a "Madame Curie strategy" first identified and described by Margaret Rossiter, who coined the phrase to describe feelings of inferiority experienced by American women scientists following Marie Curie's 1921 American tour. See Margaret W. Rossiter, *Women Scientists in America: Struggles and Strategies to 1940* (Baltimore: Johns Hopkins University Press, 1982), 130. For des Jardin's discussion of the Madame Curie Complex, see Julie des Jardins, *The Madame Curie Complex: The Hidden History of Women in Science* (New York: Feminist Press at the City University of New York, 2010), 5, 44-48.

15 There is a rich historiography on women in science and technology. Readers should consult the essays in *Removing Barriers: Women in Academic Science, Technology, Engineering, and Mathematics*, Jill M. Bystydzienski and Sharon R. Bird, eds. (Bloomington: Indiana University Press, 2006), as well as foundational works such as Rossiter, *Women Scientists in America: Struggles and Strategies to 1940*; Carolyn Merchant, *The Death of Nature: Women, Ecology and the Scientific Revolution* (New York: Harper One, 1990); Pnina G. Abir-Am and Dorinda Outram, *Uneasy Careers and Intimate Lives: Women in Science, 1789-1979* (New Brunswick, NJ: Rutgers University Press, 1987); Sandra Harding, *Whose Science? Whose Knowledge?: Thinking from Women's Lives* (Ithaca, NY: Cornell University Press,

1991); Evelyn Fox Keller, *Reflection on Gender and Science* (New Haven, Conn: Yale University Press, 1985); Keller, *Secrets of Life, Secrets of Death: Essays on Language, Gender, and Science* (New York: Routledge, 1992); Kim Tolley, *The Science Education of American Girls: A Historical Perspective* (New York: Routledge, 2003); Christine von Oertzen, *Science, Gender, and Internationalism: Women's Academic Networks, 1917-1955* (New York: Palgrave-Macmillan Press, 2012).

16 Johanna Drucker reminds humanists that "The edges and nodes that constitute the language of networks are reductive, and though they can be loaded with attributes that provide extra information and description to the depiction, they are basically static expressions of complex systems… Relationships, whether among human beings or humanistic concepts, are dynamic, fluid, flexible, and changeable. They are always in flux, not static or fixed." See Johanna Drucker, "Graphical Approaches to the Digital Humanities" in *A New Companion to the Digital Humanities*, eds. Susan Schreibman, Ray Siemans, and John Unsworth (Wiley-Blackwell: 2016), 238-250; Alberto Cairo, *The Truthful Art: Data, Charts, and Maps for Communication* (New Riders/Pearson, 2016): 14-15; Manuel Lima, *Visual Complexity: Mapping Patterns of Information* (New York: Princeton Architectural Press, 2011); Ruth Ahnert, Sebastian E. Ahnert, Catherine Nicole Coleman, and Scott B Weingart, "Visual Networks," in *The Network Turn: Changing Perspectives in the Humanities* (Cambridge: Cambridge University Press: 2020), https://www.cambridge.org/core/elements/network-turn/CC38F2EA9F51A6D1AFCB7E005218BBE5.

17 J. Stephen Catlett, *A New Guide to the Collections in the Library of the American Philosophical Society* (Philadelphia: American Philosophical Society, 1987), xviii.

18 Efforts at other institutions have suggested that public programming may be an effective way to grow collections around underrepresented groups. See, for instance, Bethany G. Anderson and Kristen Allen Wilson, "Remembering Women Scientists: The Case for Proactively Documenting Women in Science," *Collections: A Journal for Museum and Archives Professionals* 18, no. 4 (2022): 453-78.

19 "Pursuit & Persistence: 300 Years of Women in Science Opens at APS," American Philosophical Society, last modified March 30, 2023, https://www.amphilsoc.org/blog/pursuit-persistence-300-years-women-science-opens-aps. For a digital version of the exhibition, see Jane Boyd and Magdalena Hoot,

"Pursuit & Persistence: 300 Years of Women in Science (Online Exhibition)," American Philosophical Society, last modified January 11, 2024, https://www.amphilsoc.org/museum/exhibitions/pursuit-persistence.

20 Anna Doel, "Women in Science—Oral Histories from APS Members," American Philosophical Society, last modified August 26, 2024, https://www.amphilsoc.org/museum/exhibitions/women-science-oral-histories-aps-members.

21 This approach does not differ significantly from the approach used by Margaret Rossiter to document historical women of science, with the exception that our approach explicitly focuses on connections between women. As Rossiter has moved into more recent histories, her approach more closely resembles ours as well. See Margaret W. Rossiter, "The More You Look, the More You Find: Archives of Recent American Women Scientists," *Centaurus* 54, no. 4 (2012): 288-291.

22 Mushon Zer-Aviv has criticized the overuse of network models to represent all sorts of relationships, including ones that do not fit the structure of a network. See Mushon Zer-Aviv, "If everything is a network, nothing is a network," Visualizing Information for Advocacy, last modified January 8, 2016, https://visualisingadvocacy.org/node/739.html.

23 Network analysis, as a separate topic from network visualization, is a growing area of interest in the digital humanities and shows promising results for tracking larger trends in the history of science. For a robust overview of the possible uses of networks in this area, see Deryc T. Painter, Bryan C. Daniels, and Jürgen Jost, "Network Analysis for the Digital Humanities: Principles, Problems, Extensions," *Isis* 110, No. 3 (September 2019): 538-554.

24 Aaron Clauset, M. E. J. Newman, and Cristopher Moore, "Finding community structure in very large networks," *Physical Review E* 70 (December, 2004): 066111-1-066111-6, https://doi.org/10.1103/PhysRevE.70.066111.

25 Lillian Estelle Fisher, "The Influence of the Present Mexican Revolution upon the Status of Mexican Women," *The Hispanic American Historical Review* (1942): 225.

26 The suitability of finding aid data for quantitative and statistical analysis has recently been called into question, largely due to inconsistent processing and cataloguing practices. See Huw Jones and Yasmin Faghihi, "Manuscript Ca-

talogues as Data for Research: From Provenance to Data Decolonisation," *Digital Humanities Quarterly* 18, no. 3 (2024): https://www.digitalhumanities.org/dhq/vol/18/3/000751/000751.html.

27 For an overview of the application of linked data to the digital humanities, see Terhi Nurmikko-Fuller, *Linked Data for Digital Humanities* (New York: Routledge, 2024). On the intersection of uses of linked data in archives and digital humanities, see Ashleigh Hawkins, "Archives, linked data and the digital humanities: increasing access to digitised and born-digital archives via the semantic web," *Archival Science* 22 (2022): 319-344.

12

Extending the Commercial Determinants of Health: Lessons from the Opioid Industry Documents Archive

Dan Kabella, University of California, San Francisco
Kelly Ray Knight, University of California, San Francisco
Dorie E. Apollonio, University of California, San Francisco

Abstract

This paper builds on the commercial determinants of health framework through an analysis of the opioid crisis and the Opioid Industry Documents Archive (OIDA). While conceptualizations of the commercial determinants of health framework offer a critical lens on how corporate practices shape health outcomes, they have historically conceptualized corporate actors as external agents imposing harm. Drawing on insights from Science and Technology Studies, we extend existing frameworks by examining how commercial influence is embedded within and co-produced through technoscientific, regulatory, and clinical infrastructures. Using OIDA, a public repository of internal industry documents, we trace how pharmaceutical and technology companies co-constructed notions of "risk" and "responsibility" through predictive tools such as NarxCare, a clinical prescribing algorithm for opioids. This sociotechnical case study illustrates how corporate power operates through marketing or lobbying *and* through the production of knowledge, technological design, and the shaping of clinical reasoning. Our main thesis is that the commercial determinants of health framework can be extended by using the OIDA as a site to trace how corporate power is co-produced through technoscientific infrastructures that shape clinical knowledge and practice.

Introduction

A growing body of scientific research has coalesced around what Moodie et al. (2013) describe as the "science of the effects of corporate behavior on health." This emerging field interrogates the commercial systems that produce disease and reframes public health crises as outcomes of corporate forces (Jahiel and Babor 2007; Kickbusch et al. 2016; Maani et al. 2020). Grounded in frameworks like commercial determinants of health, it builds on lessons from tobacco control to conceptualize corporations as "vectors of disease" and to recast public health crises as "industrial epidemics" (Moodie et al. 2011) and "market-driven epidemics" (Rimányi et al. 2024). These paradigms invite a shift from biomedical to political-economic explanations and reflect a broader reorientation of public health science toward the analysis of policy, regulation, and economic systems.

The commercial determinants of health framework has become a key approach for understanding how corporate actors influence health outcomes at a population level. It brings attention to the ways in which behavioral, metabolic, and structural risk factors are shaped by commercial practices such as marketing, lobbying, and regulatory capture (Kickbusch et al. 2016; Freudenberg 2014). Scholars have further shown how industries intervene in science, policy, and clinical practice to protect profits and shape public discourse (Apollonio and Glantz 2017; McCoy et al. 2021). Commercial determinants of health scholars have highlighted the necessity for multisectoral strategies, the integration of both activist and scientific knowledge, and more robust safeguards against corporate influence (Kelley & Freudenberg 2022). Building on this foundation, we argue that analyzing the commercial determinants of the opioid overdose crisis requires methodological approaches that move beyond the historical framing of commercial determinants of health as an external force acting upon public health (see Figure 1). Instead, we propose understanding corporate influence as an embedded force that operates through and is constituted by unstable political, economic, technoscientific, and social infrastructures (Jasanoff 2004). This approach, consistent with modifications of the original framework published in later research (see Figure 2), opens space for a more nuanced analysis of how health is governed, produced, and contested within complex and often opaque systems of power.

Research in the field of corporate harm has been significantly shaped by analyses of industry document archives; in this paper, we focus on the Opioid Industry Documents Archive (OIDA) as our central site of analysis. Modeled after the Truth Tobacco Industry Documents, an archive resulting

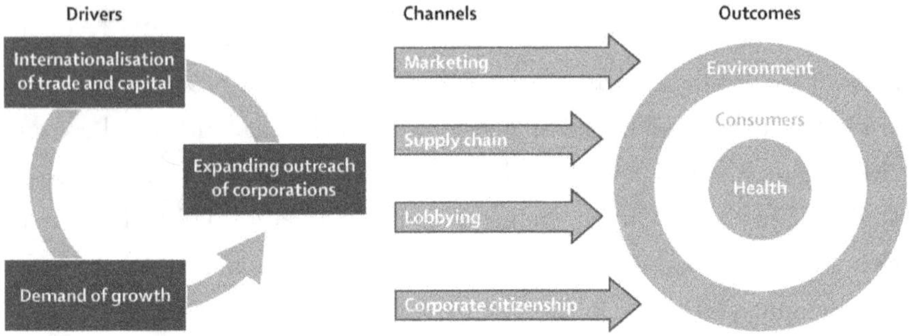

Figure 1. Original commercial determinants of health framework (2016). Source: Kickbusch I, Allen L, Franz C. The commercial determinants of health. *Lancet Glob Health.* 2016 Dec;4(12):e895-e896. doi: 10.1016/S2214-109X(16)30217-0. PMID: 27855860.

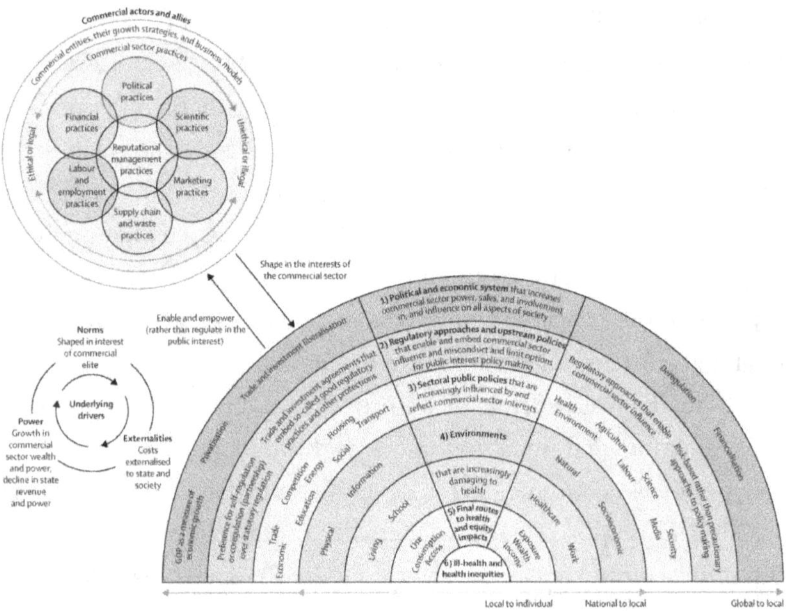

Figure 2: Modified commercial determinants of health framework (2023). Source: Gilmore AB, Fabbri A, Baum F, Bertscher A, Bondy K, Chang HJ, Demaio S, Erzse A, Freudenberg N, Friel S, Hofman KJ, Johns P, Abdool Karim S, Lacy-Nichols J, de Carvalho CMP, Marten R, McKee M, Petticrew M, Robertson L, Tangcharoensathien V, Thow AM. Defining and conceptualising the commercial determinants of health. *Lancet.* 2023 Apr 8;401(10383):1194-1213. doi: 10.1016/S0140-6736(23)00013-2. Epub 2023 Mar 23. PMID: 36966782.

from the 1998 Master Settlement Agreement in which 46 states sued major tobacco companies in order to address the costs associated with treating illnesses caused by smoking, OIDA offers unparalleled insight into how the pharmaceutical industry expanded opioid markets, obscured harm, and reshaped both pain management and addiction medicine in the United States. Rather than positioning the archive as solely a repository of corporate malfeasance, we approach it as a generative infrastructure that opens space for reimagining how knowledge about harm is produced, legitimated, and contested across institutional, community, and epistemic boundaries.

By integrating insights from science and technology studies (STS), this paper deepens the analysis of corporate influence by emphasizing the co-production of knowledge and society (Jasanoff 2004) and demonstrating how corporate actors, scientific categories, regulatory norms, values, and technologies are mutually constitutive. STS scholars have explored how epistemic authority and policy are produced together and how sociotechnical imaginaries shape public health responses (Hilgartner 2017; Murphy 2006). STS framing may illuminate the mechanisms through which corporate power becomes embedded within public health systems. It can also create space for identifying alternative epistemologies, care practices, and grassroots interventions that might challenge dominant regimes of expertise (Dumoulin et al. 2018; Mignolo 2002)

This paper challenges historical conceptions of corporate influence in science and medicine that frame it as an exogenous, disruptive force (Moodie et al. 2013). Drawing on a co-productionist framework (Jasanoff 2004), we argue that corporate power, its contestations, and alternative epistemologies are embedded within and help constitute the infrastructures of scientific knowledge, clinical practice, and public health governance. We call for an expanded approach to the archive that extends beyond industry-generated materials to include narratives of harm reduction advocates, affected patients, and grassroots organizers whose perspectives illuminate the contested terrain through which knowledge, power, and resistance circulate. Although OIDA was originally established to expose corporate malfeasance, its potential extends far beyond this function. By broadening the scope of archival inquiry to include community-based knowledge, interventions by advocates, and alternative care practices, OIDA can serve as a critical tool for understanding how corporate and state interests have become embedded within science and public health. This involves analysis that examines what corporations do and traces how systems of knowledge, regulation, and resistance are co-produced.

To illustrate this co-productive dynamic, we consider a case study of the development and deployment of health information technology systems and predictive risk scoring tools, with a focus on NarxCare (Kabella et al. 2025). Sociotechnical imaginaries, defined as "imagined forms of social life and social order that center on the development and fulfillment of innovative scientific and/or technological projects," extend beyond visions of what the future *could* look like to how they shape what kind of future society believes it *should* pursue (Jasanoff et al. 2007). In the case of NarxCare, internal documents from OIDA revealed that the platform emerged as a product of intertwined commercial, regulatory, and scientific imaginaries. Marketed to clinicians, insurers, and regulators, NarxCare employs opaque algorithms to calculate a patient's "risk" of prescription drug misuse. Pharmaceutical companies and allied tech firms played a central role in shaping this predictive infrastructure, deploying discourses of "responsible prescribing," "data-driven medicine," and "individualized risk" to reframe structural concerns about the opioid overdose crisis into a matter of patient-level surveillance and control. In doing so, they co-produced a sociotechnical regime that legitimized industry involvement in governance and displaced systemic critiques of pharmaceutical harm. This case exemplifies how sociotechnical imaginaries, when mobilized through corporate influence, shapes clinical reasoning, and public health intervention, and democratic accountability.

The Commercial Determinants of Opioid Overdose and Exposure of Corporate Harm

In her September 2024 blog post titled "Commercial interests contribute to drug use and addiction," Nora Volkow, Director of the National Institute on Drug Abuse, offered a striking shift from earlier biomedical framings of substance use disorders. She described how industries such as tobacco, cannabis, alcohol, gambling, social media, and ultra-processed food companies strategically engineer and profit from addictive behaviors and shaped environments that normalize and exacerbate substance use. Framing these forces as "commercial drivers," Volkow's post signaled an institutional acknowledgment of commercial determinants of health.

The commercial determinants of health framework was originally defined as the study of the "strategies and approaches used by the private sector to promote products and choices that are detrimental to health (Kickbusch et al. 2016)." It draws on a broad body of research in political economy, corporate power, and public health to analyze how commercial actors influence health outcomes as well as the regulatory, economic, and social environments in which populations live and work. Within this framework, corporations are understood to operate through a range of interconnected mechanisms—including marketing, lobbying, corporate social responsibility initiatives, and expansive supply chains—that together contribute to the normalization of risk and the amplification of health disparities.

Scholarship addressing commercial determinants of health has emphasized how corporate actors shape health policy, public opinion, and clinical guidelines, often through less visible or indirect means (Kickbusch et al. 2016; Maani et al. 2020). A key concept within this literature is the "corporate playbook," which refers to a recurring set of strategies used by commercial entities to advance their interests and maintain legitimacy in public relations campaigns and influencing regulatory bodies (Lacy-Nichols et al. 2022). As both an analytical and methodological tool, the corporate playbook enables researchers to track and examine patterns of influence across sectors and industries and offers a structured approach for understanding how corporate strategies are mobilized in different health contexts.

While the commercial determinants of health framework offers a critical lens on corporate influence, its analytic scope may be constrained in its original form, in which commercial actors were conceptualized as exogenous forces, rather than as entities co-constituted with the social, political, and institutional domains they inhabit. In the context of the opioid crisis, such an approach risks reducing the complex dynamics between corporate power, state mechanisms, and health systems. By framing pharmaceutical companies merely as "bad actors," this perspective overlooks the ways in which these entities have been coproduced within regulatory frameworks, pain management protocols, medical science, technology, and even public health strategies. Recognition of this weakness has led researchers to reassess directions of influence in the framework (Gilmore et al. 2023).

The Opioid Industry Documents Archive (OIDA) offers a crucial intervention into how the commercial determinants of health are understood, researched, and made actionable. As a publicly accessible digital repository co-curated by the University of California, San Francisco (UCSF) and Johns Hopkins University, OIDA houses over 26 million pages of internal corporate records released through litigation against opioid manufacturers, distributors, and consultants. These materials, including sales reports, marketing strategies, internal communications, regulatory filings, deposition transcripts, and training manuals, offer unparalleled empirical insight into the operational logics of pharmaceutical firms and the mechanisms through which they shaped the US opioid crisis.

Originally developed as part of the Global Settlement Agreement resolving lawsuits against the opioid industry, OIDA is hosted by UCSF's Library Archives and Special Collections, which also curates the Truth Tobacco Industry Documents. Building on the tobacco document archive model, UCSF expanded its industry collections over time to include records from the pharmaceutical, chemical, food, and fossil fuel sectors (OIDA 2025). These interlinked collections allow for cross-sectoral analysis of corporate influence on health and provide an evidentiary foundation for interdisciplinary scholarship spanning public health, regulatory studies, history, journalism, and digital humanities. In response to the growing recognition of corporate harm, UCSF has established dedicated centers, such as the Center for Tobacco Control Research and Education and the Center to End Corporate Harm, that focus on advancing research, training, and policy interventions aimed at addressing the pervasive impact of corporate interests on public health. This body of work underscores the urgency of critically examining the commercial determinants of health and, in the interest of this paper, within the context of the opioid crisis.

More than a static repository, OIDA constitutes a co-produced infrastructure that reconfigures the evidentiary terrain of pain management and addiction. It enables researchers, journalists, policymakers, and community advocates to move beyond abstract or deterministic models of corporate influence and instead trace the specific, situated ways in which commercial strategies were embedded within clinical practice, regulatory regimes, and public health policy. In doing so, OIDA facilitates a shift in analytical focus from individual pathology or moral failure to institutional complicity and systemic harm.

The archive's stated mission—to "shine a light" on the commercial determinants of the crisis by promoting transparency, accountability, and evidence-based research—resonates closely with Nora Volkow's call to reckon with the role of profit-driven industries in shaping patterns of addiction. While Volkow's blog post signals an epistemological shift within federal discourse, OIDA operationalizes this orientation by transforming legal evidence into public infrastructure. It invites scholars to continue extending the commercial determinants of health framework beyond a critique of external corporate interference, generating new understanding for how commercial logics become entangled with and constitutive of health governance.

From Exposures to Entanglements: Expanding the Framework

Drawing from STS, the co-productionist perspective illuminates how the pharmaceutical industry's influence in the opioid crisis took shape both through lobbying and marketing and the coordinated development of clinical science, prescribing policies, drug safety frameworks, and digital risk management technologies. For example, the creation of Risk Evaluation and Mitigation Strategies, electronic prescribing mandates, and algorithmic surveillance tools like Prescription Drug Monitoring Programs and NarxCare were regulatory responses that were shaped in tandem with evolving claims about opioid efficacy, safety, and patient risk. These systems institutionalized a policy environment that both extended the reach of pharmaceutical solutions and obscured structural causes of addiction and pain.

The concept of co-production, as developed in STS, highlights how scientific authority and social order are mutually constituted (Jasanoff 2004). Scientific knowledge does not arise in a vacuum; rather, it gains credibility and traction through alignments with institutional priorities, political logics, and cultural imaginaries. In the case of the opioid crisis, corporate actors strategically shaped the field of opioid science by funding research, partnering with professional societies, and promoting distinctions between "legitimate" and "illicit" opioid use. These efforts legitimized pharmaceutical opioid use as rational and necessary, while framing other drug use as criminal and deviant (Herzberg 2020). As Hilgartner (2017) and Murphy (2006) have shown, the authority of scientific claims is inseparable from the infrastructures through which they are assembled, circulated, and authorized.

Drawing from both analytical and pragmatic orientations of co-production (Bremer & Meisch 2017; Norström et al. 2020), we approach the OIDA as a repository of corporate malfeasance and as a site for reimagining public health knowledge-making. The archive shows that scientific practices, policy tools, and commercial interests co-evolved through iterative negotiations in which definitions of risk and standards of care were contested, revised, and stabilized. For instance, documents in the archive reveal how pharmaceutical companies worked with medical experts to redefine pain, in ways that influenced clinical guidelines and risk categorizations and made opioid prescriptions more broadly accepted and medically justified (Van Zee 2009).

Recognizing the opioid crisis as a co-produced phenomenon shifts analytical attention from discrete moments of corporate wrongdoing to the sociotechnical systems through which harm becomes routinized, normalized, and justified. For example, the opioid industry's strategic investment in the use of risk assessment tools like NarxCare, which were marketed as necessary for patient safety, helped normalize the widespread prescribing of opioids by framing it as a controlled, scientifically grounded practice, rather than corporate-driven overprescribing (Kabella et al. 2025). This perspective compels us to examine infrastructures—scientific, technological, regulatory—that shape how addiction, risk, and care are defined and managed. By treating the OIDA as a dynamic site of epistemic struggle, rather than a static record of corporate exposure, we can trace how knowledge about the opioid overdose crisis is continually co-produced. This approach reveals how corporate, regulatory, and clinical actors engage in ongoing negotiations over definitions of risk and harm. In doing so, it surfaces alternative imaginaries that contest the industry-dominated public health frameworks that have normalized and justified the opioid crisis.

Case Example: Health IT and Predictive Risk Tools

The development and implementation of Health IT and predictive risk tools like NarxCare exemplify how co-production operates in practice by materializing political rationalities, codifying contested forms of knowledge, and shaping everyday clinical encounters under the guise of neutral data-driven decision-making. Health IT constitutes a critical domain for examining how corporate power is embedded within, and operates through, both overt mechanisms of influence and the routine, encoded practices that structure everyday clinical care. NarxCare, an opioid risk prediction technology, was positioned as a neutral instrument of clinical decision-making. Closer analysis reveals it to be a sociotechnical artifact or site through which political, epistemic, and social norms are enacted and contested (Oliva 2022). Such systems encode visions of risk, responsibility, and care that reflect the interests and imaginaries of their designers, funders, and regulators.

NarxCare, now part of 52 of the 54 US prescription drug monitoring programs, illustrates how co-production operates in practice. This proprietary algorithm calculates risk scores for opioid misuse by aggregating data from sources including prescription histories, criminal justice involvement, and even consumer behavior. While marketed as a public health safeguard to prevent overprescribing, NarxCare's opaque logic, lack of FDA oversight, and unvalidated predictive methods have raised significant concerns. These concerns encompass technical flaws and regulatory gaps as well as deeper tensions around how pain management is governed and who is authorized to define legitimate care.

Drawing on internal documents from the OIDA and triangulated media sources, we found that the pharmaceutical industry actively participated in the implementation and diffusion of NarxCare (Kabella et al. 2025). Opioid companies that previously contributed to the opioid crisis sought to rebrand themselves as responsible stewards by promoting risk analytics as evidence of improved oversight. This strategic pivot shifted the framing of harm from systemic regulatory failure to the individualized mismanagement of patient risk. By embedding algorithmic risk scoring into electronic health record systems, corporate actors co-produced a new regulatory terrain: one where accountability was operationalized not through structural reform, but through intensified surveillance of patients (Chiarello 2023, Kabella et al. 2025; Oliva 2022).

NarxCare emerged through the alignment of pharmaceutical self-regulatory imperatives, emerging policy imperatives around data-driven governance, and clinical discourses about misuse and deviance. It also stabilized a broader social imaginary in which addiction is pathologized as a failure of individual responsibility, rather than a consequence of systemic and structural inequities. The result is a technological infrastructure that disciplines both patients and providers, while reinforcing boundaries between legitimate and illegitimate opioid use (Chiarello 2021; Herzberg 2020).

Public resistance to the use of NarxCare further illustrates that these infrastructures are neither total nor uncontested. A 2021 WIRED exposé revealed how chronic pain patients, especially women, were denied treatment due to algorithmic red flags (Szalavitz 2021). Grassroots advocacy groups, addiction medicine societies, and public interest law organizations have since mobilized against the use of unregulated algorithms in clinical decision-making (Buonora et al. 2024, McElfresh et al. 2023; Miotto 2022, Oliva 2022; Webster et al. 2023). Their critiques expose how algorithmic objectivity masks profound social and epistemic exclusions. These counter-narratives foreground the harms of data-driven medicine not as anomalies, but as systemic consequences of a technocratic public health regime that privileges control over care.

As such, NarxCare is more than a tool that supports clinical decisions; it remakes the terrain of health governance. It narrows the scope of care to what is legible to an algorithm, redistributes responsibility from institutions to individuals, and legitimizes surveillance under the guise of safety. It also reconfigures the meaning of medical legitimacy itself, rendering certain bodies and practices suspect by default, while insulating the institutions that produced the overdose crisis. In this way, predictive risk tools function as infrastructures of power, encoding contested moral and political judgments into the very architecture of public health.

The case of NarxCare underscores the value of co-productionist analysis in making visible the epistemic politics of health technologies. It invites us to treat tools as technical objects *and* relational actors in broader struggles over knowledge, authority, and accountability.

Conclusion

Confronting corporate power within the opioid crisis requires a critical re-examination of public health's epistemological foundations. This paper describes how corporate influence is co-produced materially and epistemically through clinical standards, regulatory frameworks, and socio-technical infrastructures. Drawing from STS, we demonstrated how tools like NarxCare illustrate the entanglement of commercial interests with health governance, creating algorithmic regimes that individualize risk and obscure structural determinants of harm. The pharmaceutical industry's role in shaping opioid science and digital surveillance infrastructures points to the need to interrogate what knowledge is produced and how that knowledge becomes institutionalized, legitimated, and contested within systems of care.

Our analysis attempts to address gaps in public health approaches, where interventions often remain anchored in behavioral and metabolic risk models, such as cessation campaigns, screenings, and health education, that may overlook the political, economic, and structural systems sustaining health inequities (Dawes 2022; Freudenberg et al. 2021; Kentikelenis and Rochford 2019; Knight 2020). We argue that integrating commercial determinants of health with tools like the Opioid Industry Documents Archive can offer new pathways for knowledge generation that address these systemic, co-constituting forces.

Acknowledgments

This research was supported by the National Institutes of Health (T32DA057216 and R01DA058687).

Bibliography

Apollonio, Dorie E., and Stanton A. Glantz. 2017. "Tobacco Industry Research on Nicotine Replacement Therapy: 'If Anyone Is Going to Take Away Our Business It Should Be Us.'" American Journal of Public Health 107, no. 10: 1636–1643. https://doi.org/10.2105/AJPH.2017.303935.

Buonora MJ, Axson SA, Cohen SM, Becker WC. Paths forward for clinicians amidst the rise of unregulated clinical decision support software: our perspective on NarxCare. Journal of general internal medicine. 2024;39(5):858-62.

Bremer, Scott, and Simon Meisch. 2017. "Co-production in climate change research: reviewing different perspectives." *Wiley Interdisciplinary Reviews: Climate Change* 8, no. 6: 482.

Chiarello, Elizabeth. 2021. "Pharmacists should treat patients who have opioid use disorders, not police them." Journal of the American Pharmacists Association 61, no. 6: e14-e19.

Chiarello, Elizabeth. 2023. "Trojan Horse technologies: smuggling criminal-legal logics into healthcare practice." American Sociological Review 88, no. 6: 1131-1160.

Dawes, Daniel E., Christian M. Amador, and Nelson J. Dunlap. "The political determinants of health: a global panacea for health inequities." In Oxford Research Encyclopedia of Global Public Health. 2022.

Dumoulin Kervran, David, Mina Kleiche-Dray, and Mathieu Quet. 2018. "Going South. How STS could think science in and with the South?." *Tapuya: Latin American Science, Technology and Society* 1, no. 1: 280-305.

Freudenberg, Nicholas, Kelley Lee, Kent Buse, Jeff Collin, Eric Crosbie, Sharon Friel, Daniel Eisenkraft Klein et al. "Defining priorities for action and research on the commercial determinants of health: a conceptual review." *American Journal of Public Health* 111, no. 12 (2021): 2202-2211.

Freudenberg, Nicholas. 2014. Lethal but Legal: Corporations, Consumption, and Protecting Public Health. Oxford: Oxford University Press.

Gilmore, Anna B., and Silvy Peeters. 2013. "Understanding corporations to inform public health policy: the example of tobacco industry interests in harm reduction and reduced risk products." *The Lancet* 382: S14.

Herzberg, David. White market drugs: Big pharma and the hidden history of addiction in America. University of Chicago Press, 2020.

Hilgartner, Stephen. 2017. Reordering life: Knowledge and control in the genomics revolution. MIT press.

Jasanoff, Sheila, Sang-Hyun Kim, and Stefan Sperling. "Sociotechnical imaginaries and science and technology policy: a cross-national comparison." NSF Research Project, Harvard University (2007).

Jasanoff, Sheila. 2024. "Ordering knowledge, ordering society." In *States of knowledge*, pp. 13-45. Routledge.

Kabella, Dan, Dorie Apollonio, Halle Young, and Kelly R. Knight. 2025. "The Rise of Clinical Decision Support Algorithms in Pain Management 2009–2024." Journal of General Internal Medicine: 1-10.

Kentikelenis, Alexander, and Connor Rochford. "Power asymmetries in global governance for health: a conceptual framework for analyzing the political-economic determinants of health inequities." *Globaliza-*

tion and Health 15 (2019): 1-10.

Kickbusch, Ilona, Luke Allen, and Christian Franz. 2016. "The commercial determinants of health." *The Lancet Global Health* 4, no. 12: 895-896.

Knight, K. R. (2020). Structural factors that affect life contexts of pregnant people with opioid use disorders: the role of structural racism and the need for structural competency. *Women's Reproductive Health*, 7(3), 164-171.

Lacy-Nichols, Jennifer, Robert Marten, Eric Crosbie, and Rob Moodie. 2022. "The public health playbook: ideas for challenging the corporate playbook." *The Lancet Global Health* 10, no. 7: e1067-e1072.

Lee, Kelley, and Nicholas Freudenberg. 2022. "Public health roles in addressing commercial determinants of health." *Annual review of public health* 43, no. 1: 375-395.

Maani, Nason, Jeff Collin, Sharon Friel, Anna B. Gilmore, Jim McCambridge, Lindsay Robertson, and Mark P. Petticrew. "Bringing the commercial determinants of health out of the shadows: a review of how the commercial determinants are represented in conceptual frameworks." *European Journal of Public Health* 30, no. 4 (2020): 660-664.

McElfresh DC, Chen L, Oliva E, et al. A call for better validation of opioid overdose risk algorithms. Journal of the American Medical Informatics Association. 2023;30(10):1741-6.

Mignolo, Walter D. 2002. "The geopolitics of knowledge and the colonial difference." *South Atlantic Quarterly* 101, no. 1: 57-96.

Miller, Clark A., and Carina Wyborn. 2020. "Co-production in global sustainability: Histories and theories." *Environmental Science & Policy* 113: 88-95.

Miotto KA. Re: Appriss NarxCare/bamboo health PMP clearinghouse implementation in California letter to Austin Weaver, Manager of California Department of Justice CURES Program. California Society of Addiction Medicine. March 2022. Available at: https://csam-asam.org/wp-531content/uploads/2022-03-28-CSAM-President-Letter-to-CURES-DOJ-re-NarxCare.pdf.532. Accessed on

August 27, 2024.

Moodie, Rob, David Stuckler, Carlos Monteiro, Nick Sheron, Bruce Neal, Thaksaphon Thamarangsi, Paul Lincoln, and Sally Casswell. 2013. "Profits and pandemics: prevention of harmful effects of tobacco, alcohol, and ultra-processed food and drink industries." *The lancet* 381, no. 9867: 670-679.

Murphy, Michelle. 2016. *Sick building syndrome and the problem of uncertainty: Environmental politics, technoscience, and women workers*. Duke University Press.

National Institute on Drug Abuse. "Commercial Interests Contribute to Drug Use and Addiction | National Institute on Drug Abuse," November 8, 2024. https://nida.nih.gov/about-nida/noras--blog/2024/09/commercial-interests-contribute-to-drug-use-addiction.

Norström, Albert V., Christopher Cvitanovic, Marie F. Löf, Simon West, Carina Wyborn, Patricia Balvanera, Angela T. Bednarek et al. 2020. "Principles for knowledge co-production in sustainability research." *Nature sustainability* 3, no. 3: 182-190.

OIDA (Opioid Industry Documents Archive). Opioid industry documents. 2025. Available at: https://www.industrydocuments.ucsf.edu/opioids. Accessed April 2, 2025.

Oliva, Jennifer D. 2022. "Dosing discrimination: regulating PDMP risk scores." Cal. L. Rev. 110: 47.

Rimányi E, Quick JD, Yamey G, Immurana M, Malik VS, Doherty T, et al. 2024. Dynamics of combatting market-driven epidemics: Insights from U.S. reduction of cigarette, sugar, and prescription opioid consumption. *PLOS Glob Public Health* 4(7): e0003479. https://doi.org/10.1371/journal.pgph.0003479

Szalavitz, M. The Pain Was Unbearable. So Why Did Doctors Turn Her Away? Wired. Aug. 11, 2021. Available at: https://www.wired.com/story/opioid-drug-addiction-algorithm-chronic-pain/. Accessed on August 01, 2024.

Van Zee, Art. 2009. "The Promotion and Marketing of OxyContin:

Commercial Triumph, Public Health Tragedy." *American Journal of Public Health* 99, no. 2: 221-227. https://doi.org/10.2105/AJPH.2007.131714.

Webster, LR and Barnes MC. CUSP FDA Citizen Petition to Protect Patients. Center for U.S. Policy, May 11, 2023. Available at: https://centerforuspolicy.org/fdacp2023-2/. Accessed on August 28, 2024.

www.ingramcontent.com/pod-product-compliance
Lightning Source LLC
LaVergne TN
LVHW022207200926
848481LV00006B/1873
9798992688832